FOCUS ON ADDITION

K-6 CONTINUITY MATHEMATICS

Coursework is available at special quantity discounts to use as premiums and sales promotions within corporate or private training programs. To obtain information or inquire about availability please write to Director, PO Box 1, Hollidaysburg, PA 16648.

NOTICE

From the US Code Collection – Violation of Copyright is a Criminal Offense

> **Criminal Infringement—Any person who infringes a copyright willfully either**
>
> (1)   for the purpose of commercial advantage or private financial gain or
>
> (2)   by the reproduction or distribution, including by electronic means during any 180-day period, of 1 or more copyrighted works, shall be punished as provided under section 2319 of title 18, United States Code.
>
> **In General.—except as otherwise provided by this title, an infringer of copyright is liable for**
>
> (1)   the copyright owner's actual damages and any additional profits of the Infringer
>
> (2)   statutory damages

Violations of copyright will be prosecuted as allowed under law.

FOCUS ON ADDITION

K-6 CONTINUITY MATHEMATICS

Note to Teaching Helper:

The first years of school are transitional years. The child will initially approach learning in the form of play and then transition to a more academic approach. As you move through the Focus On series, the instruction will begin with familiar play methods of learning and become more structured as your student masters each concept

It is important that your student's education encompass a variety of delivery methods. The best programs will include explanations of each concept, the practical worksheets presented within the textbooks, and hands on activities that encourage a deeper understanding of the ideas behind the computations.

The instruction and worksheets included in the Focus On textbooks should serve as an guide to the necessary elements of your student's mathematics program and a tool to help you determine your student's mastery of each necessary math building block. In order to ensure your student reaches their full learning potential, you should encourage hands on, creative reinforcement of each skill to complement the Focus On program. Using real life situations in tandem with the included structure and worksheets will help to fix the necessary mathematics concepts in your students mind and encourage a true understanding of the processes necessary to succeed in mathematics.

Many learners enjoy using concrete objects or tools to help them master each mathematics concept. You should work with your student to decide what concrete tools work best with their learning style. You might choose any group of pre-made math manipulatives, common household items, tally marks, or almost any item that illustrates a specific sum of objects.

As your student moves through the each level of the Focus On program it is important for you to remember mastery of each new skill is the goal. You should allow your student to take the time necessary to ensure they have gained a true understanding of the math functions presented rather than focusing on completing the program during a pre-set term. Statistics have shown that a child will succeed at mathematics if they gain mastery of each concept before attempting to learn subsequent concepts. You should use the included guide and worksheets to assess your child's progress and only move on to the next segment once you are certain that your student has achieved mastery

The proven path to mastery in mathematics is practice, practice, practice. This does not mean that you should present mathematics in the form of mindless repetition. You should present varied opportunities to approach each mathematics concept from a variety of angles. You should also encourage reinforcement of each previously mastered concept along with the presentation of each new concept. This allows the child to build upon a solid foundation of comprehension that will encourage math mastery throughout their educational career.

The focus on series of lessons bring a laser focus to each subject area, ensuring each necessary skill is fully developed before the student moves onto the next skill in the series.

Continuity is critical to educational success. By ensuring your student gains a comprehensive understanding of each skill before transitioning to the next skill, you lay the groundwork for lifelong academic success.

Unit 1 - Counting to Add

Instructor Note:

Your student should gain a basic understanding of the concept that addition increases the final number.

You can begin the process of teaching counting to add by using concrete objects to illustrate how two groups of items can be combined to form a larger group.

As your student gains the ability to add concrete objects, you can transition these skills to counting pictures in worksheets.

1.   X the box that shows one more sun.

2.   X the box that shows one more ball.

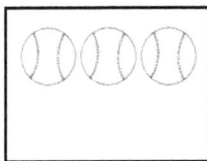

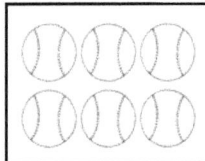

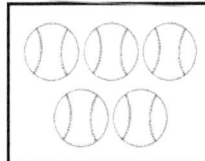

3. X the box that shows one more crayon.

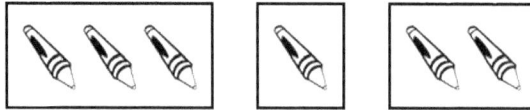

4. X the box that shows one more apple.

5. X the box that shows one more frog.

Name: _____

1. X the box that shows 2 more elephants.

2. X the box that shows 2 more fish.

3. X the box that shows 2 more flags.

4. X the box that shows 2 more pigs.

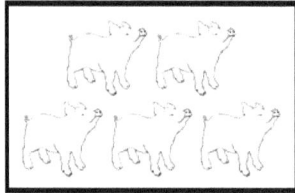

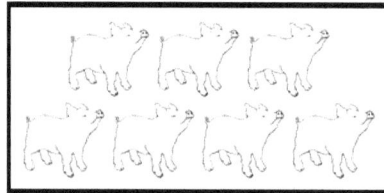

5. X the box that shows 2 more butterflies.

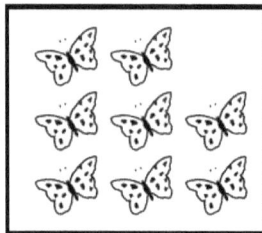

6. X the box that shows 2 more pumpkins.

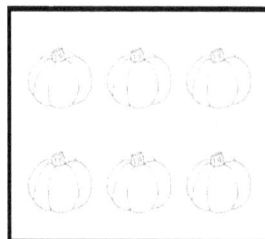

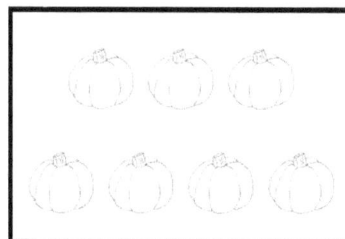

4

Name: _____

1.

X the box that shows 3 more hats.

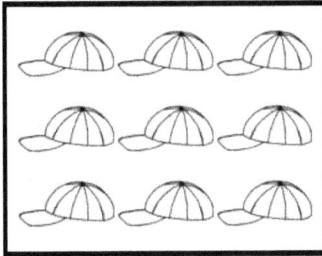

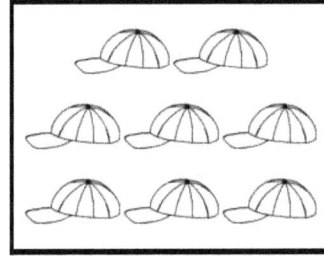

2.

X the box that shows 3 more birds.

3.

X the box that shows 3 more triangles.

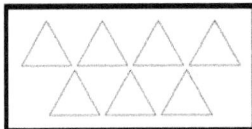

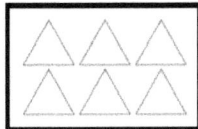

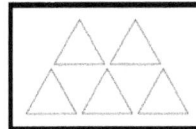

4. X the box that shows 3 more alligators.

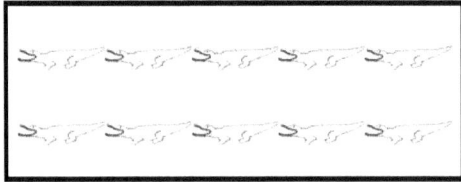

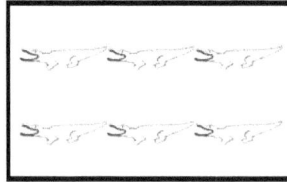

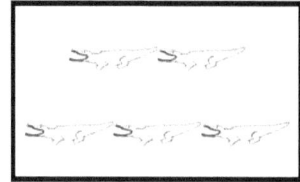

5. X the box that shows 3 more jets.

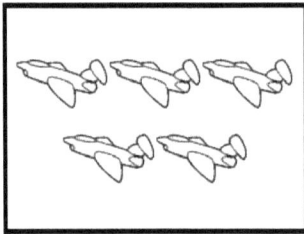

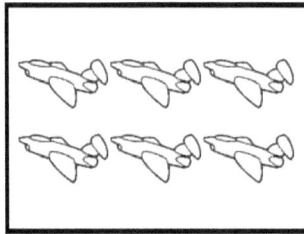

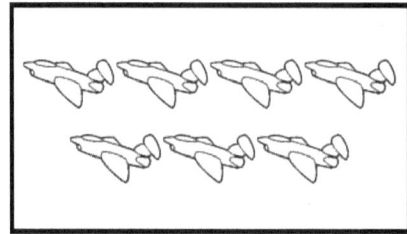

6. X the box that shows 3 more cakes.

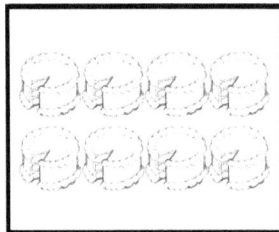

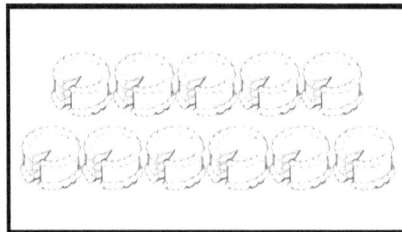

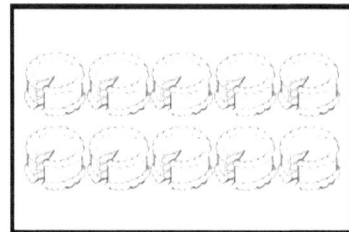

Name: _____

1. Draw 4 more balls. Write how many balls you have now.

2. Draw 4 more triangles. Write how many triangles you have now.

3. Draw 4 more arrows. Write how many arrows you have now.

4. Draw 4 more circles. Write how many circles you have now.

5. Draw 4 more hats. Write how many hats you have now.

Name: _____

1. Draw 5 more cans. Write how many cans you have now.

2. Draw 5 more cats. Write how many cats you have now.

3. Draw 5 more drums. Write how many drums you have now.

4. Draw 5 more flowers. Write how many flowers you have now.

5. Draw 5 more rectangles. Write how many rectangles you have now.

Name: _____

1. Draw 6 more nuts. Write how many nuts you have now.

_____

2. Draw 6 more pans. Write how many pans you have now...

_____

3. Draw 6 more squares. Write how many squares you have now.

_____

4. Draw 6 more trees. Write how many trees you have now.

_____

5. Draw 6 more tubs. Write how many tubs you have now.

_____

Unit 2 - Model Addition

Instructor Note:

Use models to apply basic addition facts.

Student Instruction:

When you add two groups of objects together, they make a larger group.

If you count the objects in each group and make them one large group, you are adding.

3 Stars   3 Stars   6 Stars in All

1.   Add the drums and then X the box that shows the right answer.

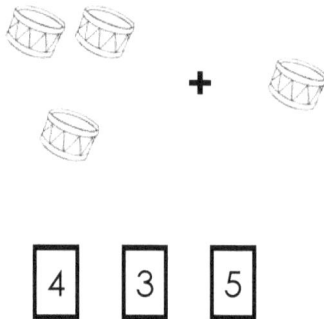

| | 4 | 3 | 5 |

___

2.   Add the bats and then X the box that shows the right answer.

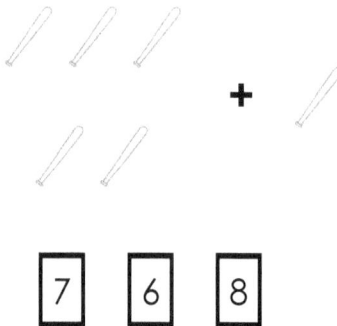

| 7 | 6 | 8 |

3. Add the cans and then X the box that shows the right answer.

☐☐ + ☐

| 4 | 3 | 5 |

---

4. Add the cats and then X the box that shows the right answer.

+

| 7 | 10 | 8 |

---

5. Add the hens and then X the box that shows the right answer.

+

| 4 | 7 | 5 |

11

Name: _____

1.   Add the circles and then X the box that shows the right answer.

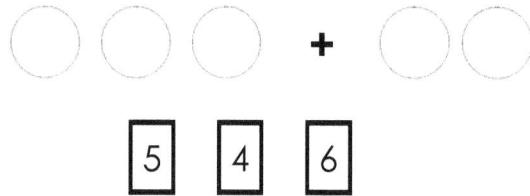

○ ○ ○ **+** ○ ○

| 5 | | 4 | | 6 |

---

2.   Add the clowns and then X the box that shows the right answer.

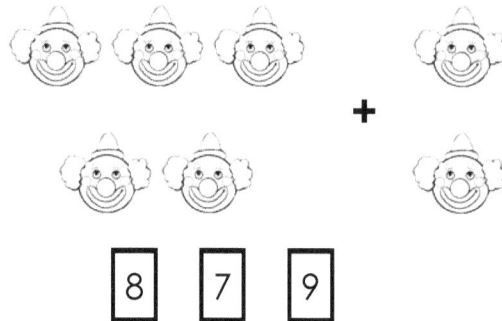

+

| 8 | | 7 | | 9 |

---

3.   Add the crabs and then X the box that shows the right answer.

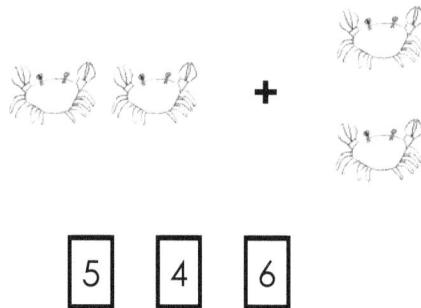

+

| 5 | | 4 | | 6 |

---

4. Add the dogs and then X the box that shows the right answer.

+

| 8 | 10 | 9 |

---

5. Add the elephants and then X the box that shows the right answer.

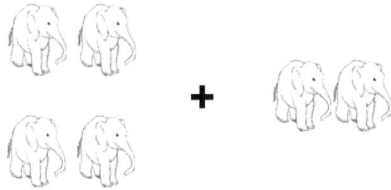

+

| 5 | 4 | 6 |

---

1. Add the fish and then X the box that shows the right answer.

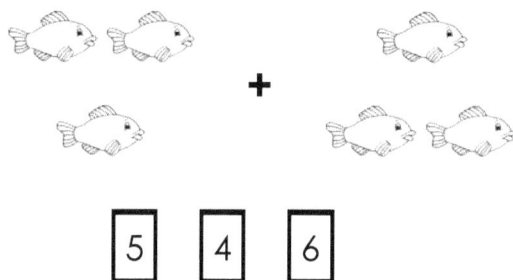

    +

    | 5 | | 4 | | 6 |

---

2. Add the flags and then X the box that shows the right answer.

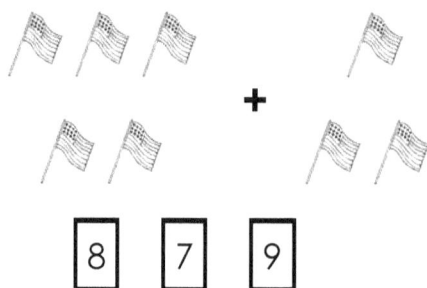

    +

    | 8 | | 7 | | 9 |

---

3. Add the flowers and then X the box that shows the right answer.

    +

    | 5 | | 4 | | 6 |

---

4. Add the gifts and then X the box that shows the right answer.

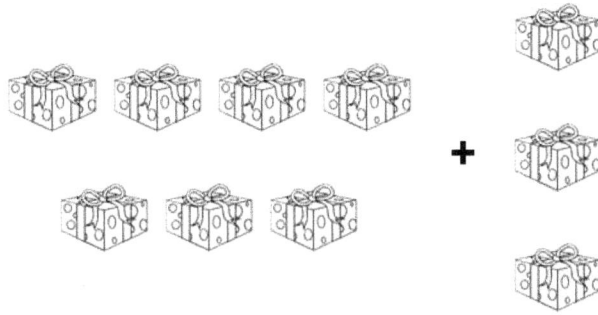

| 8 | 10 | 9 |

---

5. Add the hats and then X the box that shows the right answer.

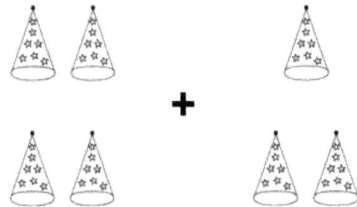

| 5 | 7 | 6 |

Name: _____

1. Add the pigs and then X the box that shows the right answer.

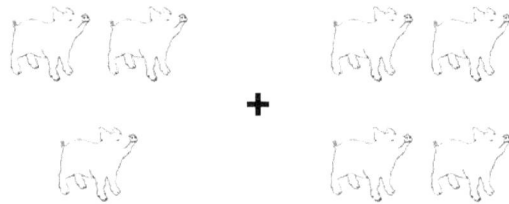

+

| 7 | 8 | 6 |

2. Add the penguins and then X the box that shows the right answer.

+

| 8 | 7 | 9 |

3. Add the porcupines and then X the box that shows the right answer.

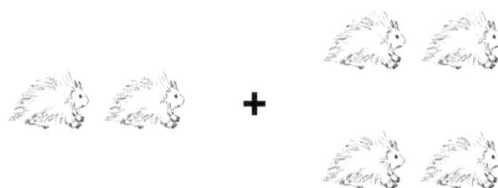

+

| 5 | 4 | 6 |

4. Add the beds and then X the box that shows the right answer.

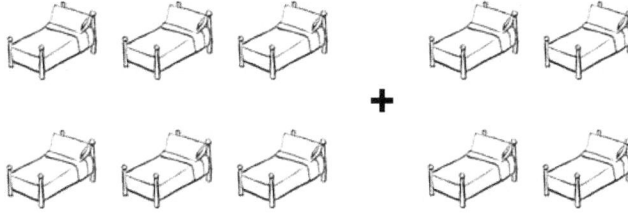

8	10	9

---

5. Add the eels and then X the box that shows the right answer.

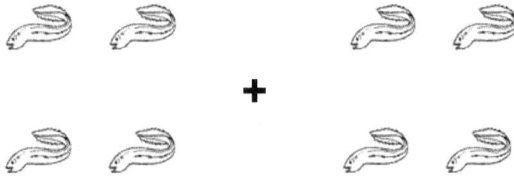

7	8	9

Name: _____

1. Add the lions and then X the box that shows the right answer.

+

| 7 | 8 | 6 |

2. Add the nuts and then X the box that shows the right answer.

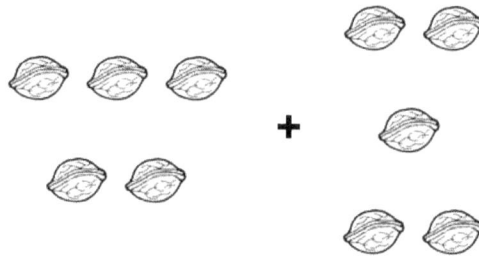

+

| 8 | 10 | 9 |

3. Add the nets and then X the box that shows the right answer.

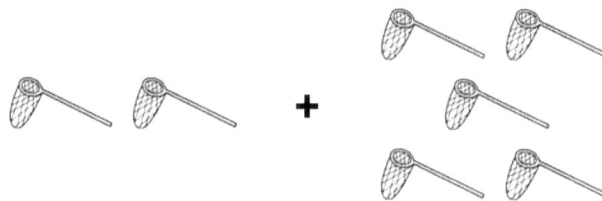

+

| 5 | 7 | 6 |

4. Add the pans and then X the box that shows the right answer.

$$+$$

| 11 | | 10 | | 12 |

---

5. Add the rats and then X the box that shows the right answer.

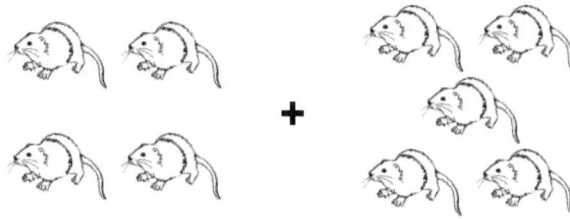

$$+$$

| 7 | 8 | 9 |

Name: _____

1. Add the rectangles and then X the box that shows the right answer.

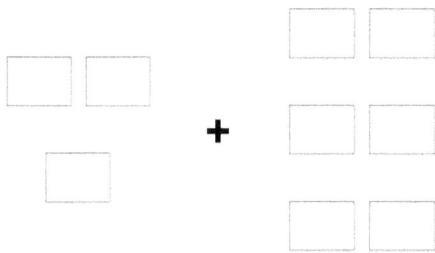

+

| 7 | 8 | 9 |

2. Add the trains and then X the box that shows the right answer.

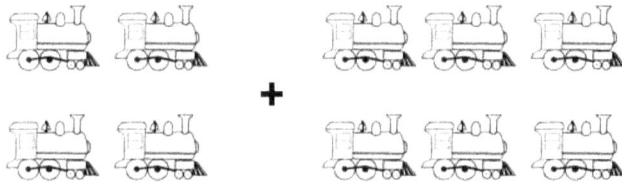

+

| 11 | 10 | 9 |

3. Add the bins and then X the box that shows the right answer.

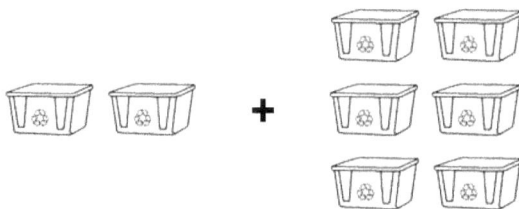

+

| 8 | 9 | 7 |

4. Add the hands and then X the box that shows the right answer.

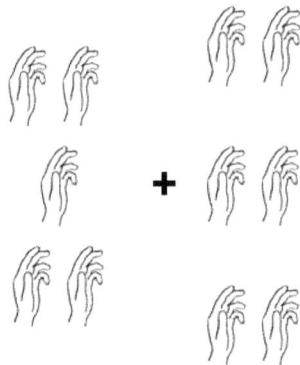

$$+$$

| 11 | 10 | 9 |

5. Add the sharks and then X the box that shows the right answer.

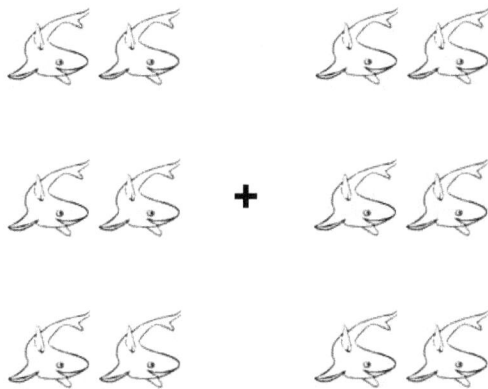

$$+$$

| 11 | 10 | 12 |

1. Add the goats and then X the box that shows the right answer.

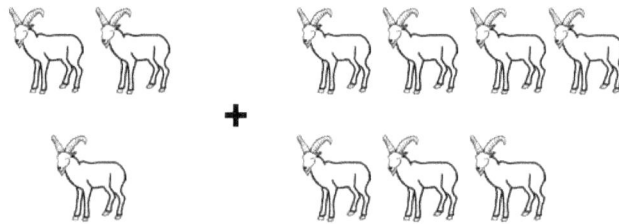

| 11 | 9 | 10 |

2. Add the squares and then X the box that shows the right answer.

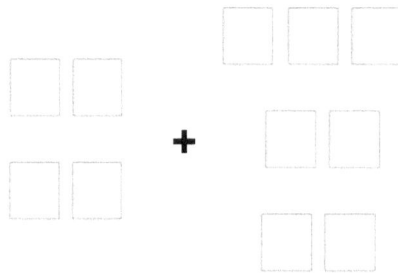

| 11 | 10 | 12 |

3. Add the pens and then X the box that shows the right answer.

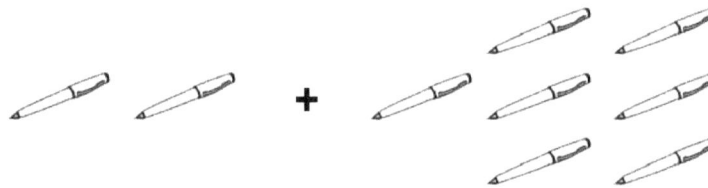

| 8 | 9 | 7 |

4. Add the tents and then X the box that shows the right answer.

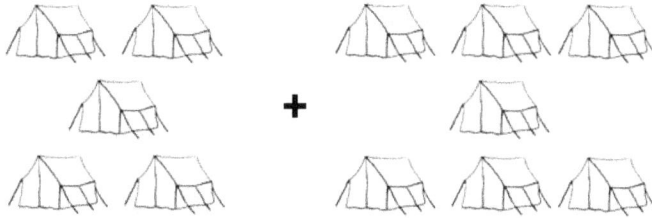

| 11 | 10 | 12 |

5. Add the bears and then X the box that shows the right answer.

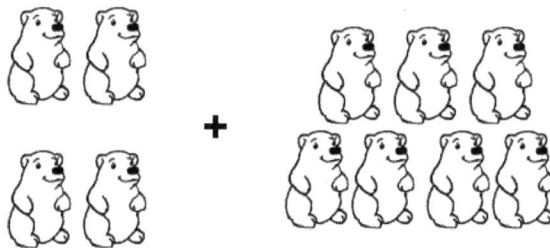

| 11 | 13 | 12 |

1. Add the trees and then X the box that shows the right answer.

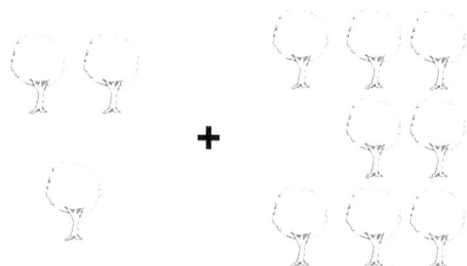

| 11 | 9 | 10 |

2. Add the whales and then X the box that shows the right answer.

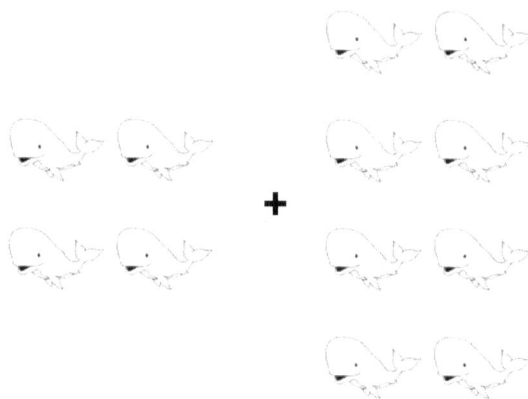

| 11 | 10 | 12 |

3.   Add the logs and then X the box that shows the right answer.

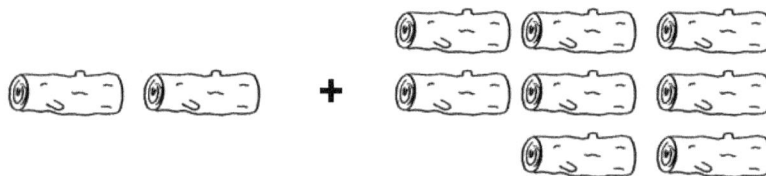

$$\boxed{11} \quad \boxed{9} \quad \boxed{10}$$

---

4.   Add the pins and then X the box that shows the right answer.

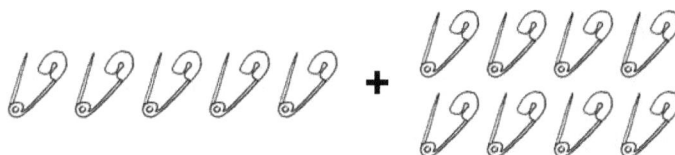

$$\boxed{11} \quad \boxed{13} \quad \boxed{12}$$

---

5.   Add the tubs and then X the box that shows the right answer.

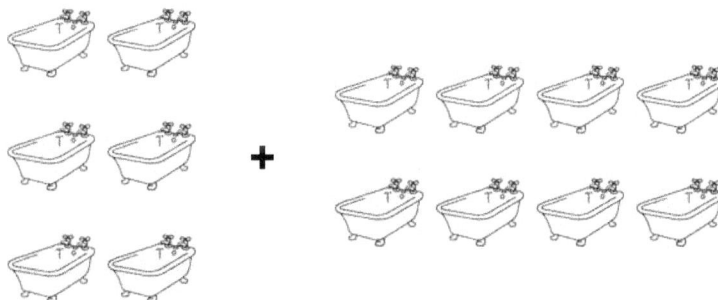

$$\boxed{14} \quad \boxed{13} \quad \boxed{12}$$

---

1. Add the cakes and then X the box that shows the right answer.

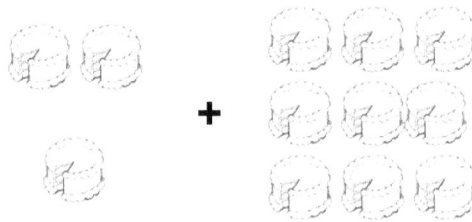

+

| 11 | | 12 | | 10 |

---

2. Add the triangles and then X the box that shows the right answer.

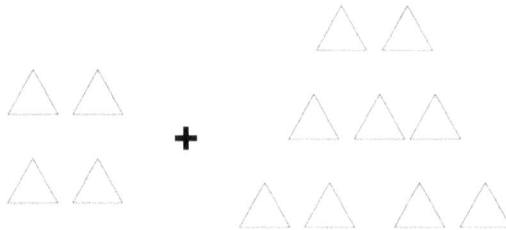

+

| 11 | | 13 | | 12 |

---

3. Add the fish and then X the box that shows the right answer.

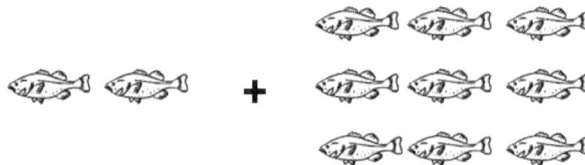

+

| 11 | | 9 | | 10 |

---

4. Add the balls and then X the box that shows the right answer.

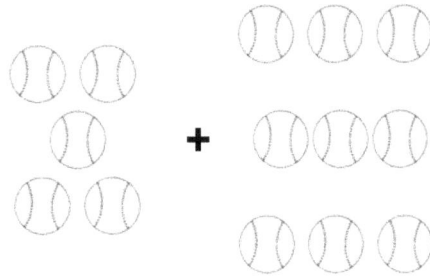

+

14   13   12

---

5. Add the crayons and then X the box that shows the right answer.

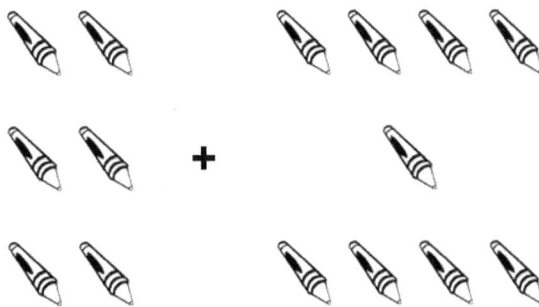

+

14   13   15

Unit 3 - Addition Using a Number Line

Instructor Note:

Your student should gain the ability to use a number line in place of concrete objects when solving addition problems.

Student Instruction:

Number lines can help you add. The number line will have a beginning number and an ending number.

This number line begins at the number 0 and ends at the number 24. That means that you can add any numbers whose answer is 0 and higher or 24 and lower.

Use this number line to add 7 + 5

Place your pencil in the number 7.

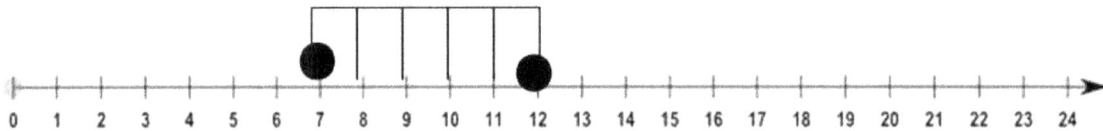

Jump your pencil forward 5 places landing on 8, 9, 10, 11, and 12. The last number you land on is the answer.

7 + 5 = 12

Use the number line to solve each problem.

1.

Use the number line below to solve 4 + 4 = _____

2.

Use the number line below to solve 18 + 3 = _____

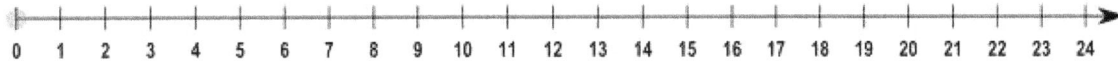

3.

Use the number line below to solve 10 + 5 = _____

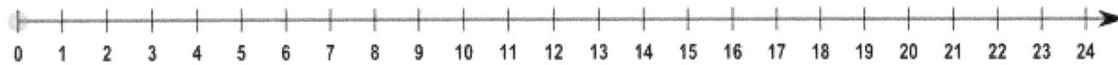

4.

Use the number line below to solve 6 + 6 = _____

5.

Use the number line below to solve 20 + 3 = _____

Name: _____

Use the number line to solve each problem.

1.

Use the number line below to solve 6 + 9 = _____

2.

Use the number line below to solve 8 + 8 = _____

3.

Use the number line below to solve 7 + 5 = _____

4.

Use the number line below to solve 6 + 3 = _____

Name: _____

Use the number line to solve each problem.

1.

Use the number line below to solve 9 + 5 = _____

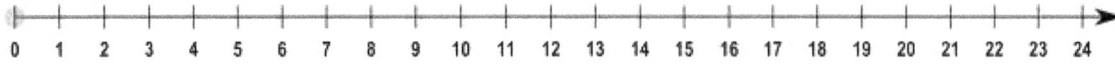

2.

Use the number line below to solve 7 + 4 = _____

3.

Use the number line below to solve 4 + 3 = _____

4.

Use the number line below to solve 6 + 2 = _____

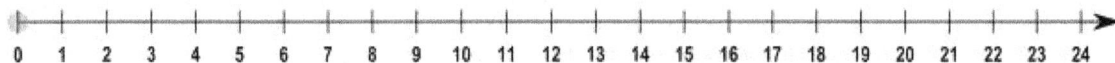

Name: _____

Use the number line to solve each problem.

1.

Use the number line below to solve 4 + 2 = _____

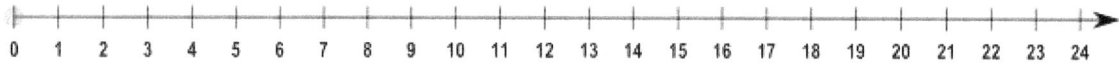

2.

Use the number line below to solve 7 + 3 = _____

3.

Use the number line below to solve 9 + 4 = _____

4.

Use the number line below to solve 5 + 5 = _____

Name: _____

Use the number line to solve each problem.

1.

Use the number line below to solve 2 + 6 = _____

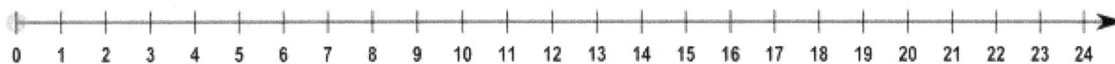

```
0  1  2  3  4  5  6  7  8  9  10 11 12 13 14 15 16 17 18 19 20 21 22 23 24
```

2.

Use the number line below to solve 3 + 7 = _____

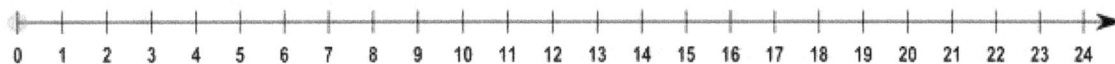

```
0  1  2  3  4  5  6  7  8  9  10 11 12 13 14 15 16 17 18 19 20 21 22 23 24
```

3.

Use the number line below to solve 6 + 8 = _____

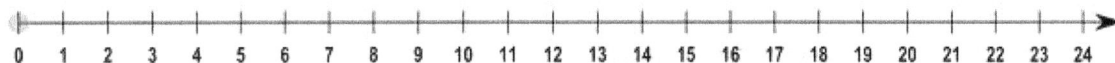

```
0  1  2  3  4  5  6  7  8  9  10 11 12 13 14 15 16 17 18 19 20 21 22 23 24
```

4.

Use the number line below to solve 4 + 9 = _____

```
0  1  2  3  4  5  6  7  8  9  10 11 12 13 14 15 16 17 18 19 20 21 22 23 24
```

Name: _____

Use the number line to solve each problem.

1.

Use the number line below to solve 4 + 8 = _____

```
+--+--+--+--+--+--+--+--+--+--+--+--+--+--+--+--+--+--+--+--+--+--+--+--+-->
0  1  2  3  4  5  6  7  8  9  10 11 12 13 14 15 16 17 18 19 20 21 22 23 24
```

2.

Use the number line below to solve 9 + 9 = _____

```
+--+--+--+--+--+--+--+--+--+--+--+--+--+--+--+--+--+--+--+--+--+--+--+--+-->
0  1  2  3  4  5  6  7  8  9  10 11 12 13 14 15 16 17 18 19 20 21 22 23 24
```

3.

Use the number line below to solve 2 + 7 = _____

```
+--+--+--+--+--+--+--+--+--+--+--+--+--+--+--+--+--+--+--+--+--+--+--+--+-->
0  1  2  3  4  5  6  7  8  9  10 11 12 13 14 15 16 17 18 19 20 21 22 23 24
```

4.

Use the number line below to solve 8 + 6 = _____

```
+--+--+--+--+--+--+--+--+--+--+--+--+--+--+--+--+--+--+--+--+--+--+--+--+-->
0  1  2  3  4  5  6  7  8  9  10 11 12 13 14 15 16 17 18 19 20 21 22 23 24
```

Name: _____

Use the number line to solve each problem.

1.

Use the number line below to solve 6 + 5 = _____

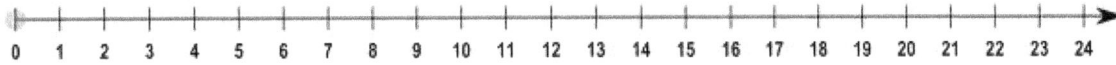

2.

Use the number line below to solve 4 + 4 = _____

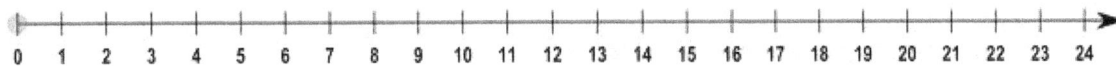

3.

Use the number line below to solve 9 + 3 = _____

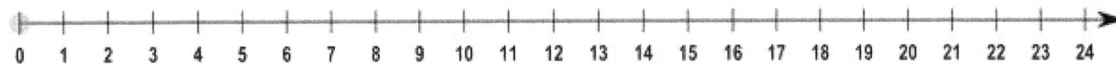

4.

Use the number line below to solve 5 + 2 = _____

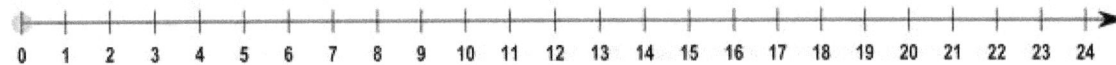

Name: _____

Use the number line to solve each problem.

1.

Use the number line below to solve 9 + 1 = _____

2.

Use the number line below to solve 8 + 2 = _____

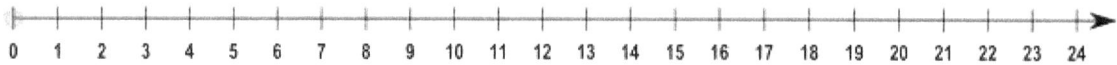

3.

Use the number line below to solve 8 + 4 = _____

4.

Use the number line below to solve 6 + 3 = _____

Use the number line to solve each problem.

1.

Use the number line below to solve 9 + 6 = _____

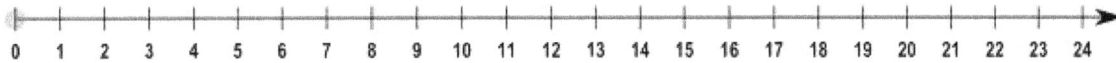

```
+--+--+--+--+--+--+--+--+--+--+--+--+--+--+--+--+--+--+--+--+--+--+--+--+-->
0  1  2  3  4  5  6  7  8  9  10 11 12 13 14 15 16 17 18 19 20 21 22 23 24
```

2.

Use the number line below to solve 7 + 7 = _____

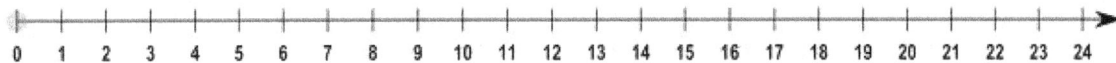

```
+--+--+--+--+--+--+--+--+--+--+--+--+--+--+--+--+--+--+--+--+--+--+--+--+-->
0  1  2  3  4  5  6  7  8  9  10 11 12 13 14 15 16 17 18 19 20 21 22 23 24
```

3.

Use the number line below to solve 2 + 8 = _____

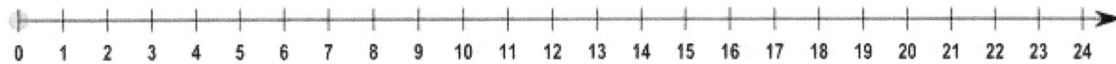

```
+--+--+--+--+--+--+--+--+--+--+--+--+--+--+--+--+--+--+--+--+--+--+--+--+-->
0  1  2  3  4  5  6  7  8  9  10 11 12 13 14 15 16 17 18 19 20 21 22 23 24
```

4.

Use the number line below to solve 5 + 9 = _____

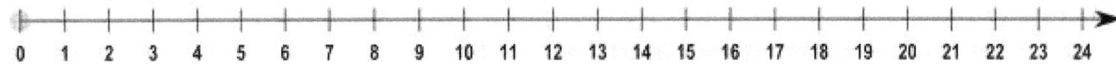

```
+--+--+--+--+--+--+--+--+--+--+--+--+--+--+--+--+--+--+--+--+--+--+--+--+-->
0  1  2  3  4  5  6  7  8  9  10 11 12 13 14 15 16 17 18 19 20 21 22 23 24
```

Name: _____

Use the number line to solve each problem.

1.

Use the number line below to solve 8 + 9 = _____

2.

Use the number line below to solve 3 + 8 = _____

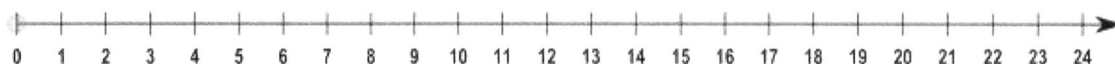

3.

Use the number line below to solve 6 + 7 = _____

4.

Use the number line below to solve 4 + 6 = _____

Unit 4 - Addition - Facts

Instructor Note:

Your student should memorize the basic addition facts so that they can add from memory instead of using tools. The addition fact chart and flash cards are excellent tools to assist you in helping your student master basic addition facts. You should also copy the pages from the unit transitioning to mental math and have your student complete these problems until they have memorized each addition fact.

Student Instruction:

Adding numbers is much easier if you memorize some addition facts. Knowing the answers to basic facts will help you to solve addition problems much more quickly.

Cut out the flash cards and practice your addition facts.

1 + 0	1 + 1	1 + 2
1 + 3	1 + 4	1 + 5
1 + 6	1 + 7	1 + 8
1 + 9	2 + 0	2 + 1

2 + 2	2 + 3	2 + 4
2 + 5	2 + 6	2 + 7
2 + 8	2 + 9	3 + 0
3 + 1	3 + 2	3 + 3
3 + 4	3 + 5	3 + 6
3 + 7	3 + 8	3 + 9
4 + 0	4 + 1	4 + 2

4 + 3	4 + 4	4 + 5
4 + 6	4 + 7	4 + 8
4 + 9	5 + 0	5 + 1
5 + 2	5 + 3	5 + 4
5 + 5	5 + 6	5 + 7
5 + 8	5 + 9	6 + 0
6 + 1	6 + 2	6 + 3

6 + 4	6 + 5	6 + 6
6 + 7	6 + 8	6 + 9
7 + 0	7 + 1	7 + 2
7 + 3	7 + 4	7 + 5
7 + 6	7 + 7	7 + 8
7 + 9	8 + 0	8 + 1
8 + 2	8 + 3	8 + 4

8 + 5	8 + 6	8 + 7
8 + 8	8 + 9	9 + 0
9 + 1	9 + 2	9 + 3
9 + 4	9 + 5	9 + 6
9 + 7	9 + 8	9 + 9

Addition Facts

Try It – Complete the addition facts chart.

1 +0	1 +1	1 +2	1 +3	1 +4	1 +5	1 +6	1 +7	1 +8	1 +9
2 +0	2 +1	2 +2	2 +3	2 +4	2 +5	2 +6	2 +7	2 +8	2 +9
3 +0	3 +1	3 +2	3 +3	3 +4	3 +5	3 +6	3 +7	3 +8	3 +9
4 +0	4 +1	4 +2	4 +3	4 +4	4 +5	4 +6	4 +7	4 +8	4 +9
5 +0	5 +1	5 +2	5 +3	5 +4	5 +5	5 +6	5 +7	5 +8	5 +9
6 +0	6 +1	6 +2	6 +3	6 +4	6 +5	6 +6	6 +7	6 +8	6 +9
7 +0	7 +1	7 +2	7 +3	7 +4	7 +5	7 +6	7 +7	7 +8	7 +9
8 +0	8 +1	8 +2	8 +3	8 +4	8 +5	8 +6	8 +7	8 +8	8 +9
9 +0	9 +1	9 +2	9 +3	9 +4	9 +5	9 +6	9 +7	9 +8	9 +9

Unit 5 - Addition - Transitioning to Mental Math

Instructor Note:

Your student should begin to transition addition using concrete object modeling to number only addition. You can help them accomplish this transition by providing your student with addition problems with and then without visual aides.

Your student should move through the stages of completing addition with concrete objects, worksheet based concrete object aids, and then with no visual aids whatsoever.

Student Instructions:

When you add two groups of objects together, they form a larger group.

You can count numbers in your head the same way that you count pictures.

Example: 2 + 3 = 5

Start at the number 2.

Hold the number 2 in your head.

Count the next 3 numbers aloud.

3, 4, 5

The last number you count is 5, so 2 + 3 = 5.

Name: _____

Add the numbers and then pick the box with the correct answer to each problem.

1.
$$\begin{array}{r} 9 \\ +\,0 \\ \hline \end{array}$$

| 10 | 0 | 9 |

2.
$$\begin{array}{r} 5 \\ +\,0 \\ \hline \end{array}$$

| 0 | 4 | 5 |

3.
$$\begin{array}{r} 7 \\ +\,0 \\ \hline \end{array}$$

| 6 | 0 | 7 |

4.
$$\begin{array}{r} 2 \\ +\,0 \\ \hline \end{array}$$

| 2 | 3 | 0 |

5.
$$\begin{array}{r} 4 \\ +\,0 \\ \hline \end{array}$$

| 4 | 5 | 0 |

6.
$$\begin{array}{r} 3 \\ +\,0 \\ \hline \end{array}$$

| 4 | 3 | 0 |

7.
$$\begin{array}{r} 6 \\ +\,0 \\ \hline \end{array}$$

| 5 | 6 | 0 |

8.
$$\begin{array}{r} 8 \\ +\,0 \\ \hline \end{array}$$

| 9 | 8 | 0 |

Name: _____

Add the numbers and then pick the box with the correct answer to each problem.

1.
$$\begin{array}{r} 9 \\ +\ 1 \\ \hline \end{array}$$

| 11 | | 8 | | 10 |

2.
$$\begin{array}{r} 5 \\ +\ 1 \\ \hline \end{array}$$

| 3 | | 4 | | 6 |

3.
$$\begin{array}{r} 7 \\ +\ 1 \\ \hline \end{array}$$

| 7 | | 6 | | 8 |

4.
$$\begin{array}{r} 2 \\ +\ 1 \\ \hline \end{array}$$

| 3 | | 4 | | 1 |

5.
$$\begin{array}{r} 4 \\ +\ 1 \\ \hline \end{array}$$

| 5 | | 6 | | 7 |

6.
$$\begin{array}{r} 3 \\ +\ 1 \\ \hline \end{array}$$

| 5 | | 4 | | 6 |

7.
$$\begin{array}{r} 6 \\ +\ 1 \\ \hline \end{array}$$

| 5 | | 6 | | 7 |

8.
$$\begin{array}{r} 8 \\ +\ 1 \\ \hline \end{array}$$

| 9 | | 7 | | 10 |

Name: _____

Add the numbers and then pick the box with the correct answer to each problem.

1.
$$\begin{array}{r} 9 \\ +2 \\ \hline \end{array}$$

| 12 | 9 | 11 |

2.
$$\begin{array}{r} 5 \\ +2 \\ \hline \end{array}$$

| 4 | 5 | 7 |

3.
$$\begin{array}{r} 7 \\ +2 \\ \hline \end{array}$$

| 8 | 6 | 9 |

4.
$$\begin{array}{r} 2 \\ +2 \\ \hline \end{array}$$

| 3 | 4 | 5 |

5.
$$\begin{array}{r} 4 \\ +2 \\ \hline \end{array}$$

| 5 | 6 | 7 |

6.
$$\begin{array}{r} 3 \\ +2 \\ \hline \end{array}$$

| 5 | 4 | 6 |

7.
$$\begin{array}{r} 6 \\ +2 \\ \hline \end{array}$$

| 8 | 6 | 7 |

8.
$$\begin{array}{r} 8 \\ +2 \\ \hline \end{array}$$

| 10 | 9 | 7 |

Name: _____

Add the numbers and then pick the box with the correct answer to each problem.

1.
$$\begin{array}{r} 9 \\ +3 \\ \hline \end{array}$$

| 13 | | 10 | | 12 |

2.
$$\begin{array}{r} 5 \\ +3 \\ \hline \end{array}$$

| 7 | | 6 | | 8 |

3.
$$\begin{array}{r} 7 \\ +3 \\ \hline \end{array}$$

| 11 | | 10 | | 9 |

4.
$$\begin{array}{r} 2 \\ +3 \\ \hline \end{array}$$

| 6 | | 4 | | 5 |

5.
$$\begin{array}{r} 4 \\ +3 \\ \hline \end{array}$$

| 8 | | 6 | | 7 |

6.
$$\begin{array}{r} 3 \\ +3 \\ \hline \end{array}$$

| 5 | | 4 | | 6 |

7.
$$\begin{array}{r} 6 \\ +3 \\ \hline \end{array}$$

| 8 | | 9 | | 7 |

8.
$$\begin{array}{r} 8 \\ +3 \\ \hline \end{array}$$

| 10 | | 9 | | 11 |

Name: _____

Add the numbers and then pick the box with the correct answer to each problem.

1.
$$\begin{array}{r} 9 \\ +\ 4 \\ \hline \end{array}$$

2.
$$\begin{array}{r} 5 \\ +\ 4 \\ \hline \end{array}$$

| 14 | 11 | 13 |

| 8 | 7 | 9 |

3.
$$\begin{array}{r} 7 \\ +\ 4 \\ \hline \end{array}$$

4.
$$\begin{array}{r} 2 \\ +\ 4 \\ \hline \end{array}$$

| 12 | 11 | 10 |

| 7 | 5 | 6 |

5.
$$\begin{array}{r} 4 \\ +\ 4 \\ \hline \end{array}$$

6.
$$\begin{array}{r} 3 \\ +\ 4 \\ \hline \end{array}$$

| 9 | 7 | 8 |

| 8 | 9 | 7 |

7.
$$\begin{array}{r} 6 \\ +\ 4 \\ \hline \end{array}$$

8.
$$\begin{array}{r} 8 \\ +\ 4 \\ \hline \end{array}$$

| 9 | 10 | 11 |

| 11 | 13 | 12 |

Name: _____

Add the numbers and then pick the box with the correct answer to each problem.

1.
$$\begin{array}{r} 9 \\ +5 \\ \hline \end{array}$$

| 15 | 12 | 14 |

2.
$$\begin{array}{r} 5 \\ +5 \\ \hline \end{array}$$

| 9 | 11 | 10 |

3.
$$\begin{array}{r} 7 \\ +5 \\ \hline \end{array}$$

| 13 | 12 | 11 |

4.
$$\begin{array}{r} 2 \\ +5 \\ \hline \end{array}$$

| 8 | 9 | 7 |

5.
$$\begin{array}{r} 4 \\ +5 \\ \hline \end{array}$$

| 10 | 11 | 9 |

6.
$$\begin{array}{r} 3 \\ +5 \\ \hline \end{array}$$

| 8 | 9 | 7 |

7.
$$\begin{array}{r} 6 \\ +5 \\ \hline \end{array}$$

| 10 | 11 | 12 |

8.
$$\begin{array}{r} 8 \\ +5 \\ \hline \end{array}$$

| 12 | 14 | 13 |

Name: _____

Add the numbers and then pick the box with the correct answer to each problem.

1.
$$\begin{array}{r} 9 \\ + 6 \\ \hline \end{array}$$

| 16 | | 13 | | 15 |

2.
$$\begin{array}{r} 5 \\ + 6 \\ \hline \end{array}$$

| 10 | | 12 | | 11 |

3.
$$\begin{array}{r} 7 \\ + 6 \\ \hline \end{array}$$

| 14 | | 13 | | 12 |

4.
$$\begin{array}{r} 2 \\ + 6 \\ \hline \end{array}$$

| 9 | | 10 | | 8 |

5.
$$\begin{array}{r} 4 \\ + 6 \\ \hline \end{array}$$

| 11 | | 12 | | 10 |

6.
$$\begin{array}{r} 3 \\ + 6 \\ \hline \end{array}$$

| 9 | | 10 | | 8 |

7.
$$\begin{array}{r} 6 \\ + 6 \\ \hline \end{array}$$

| 11 | | 12 | | 13 |

8.
$$\begin{array}{r} 8 \\ + 6 \\ \hline \end{array}$$

| 13 | | 15 | | 14 |

Name: _____

Add the numbers and then pick the box with the correct answer to each problem.

1.
$$\begin{array}{r} 9 \\ +7 \\ \hline \end{array}$$

| 17 | | 14 | | 16 |

2.
$$\begin{array}{r} 5 \\ +7 \\ \hline \end{array}$$

| 11 | | 13 | | 12 |

3.
$$\begin{array}{r} 7 \\ +7 \\ \hline \end{array}$$

| 15 | | 14 | | 13 |

4.
$$\begin{array}{r} 2 \\ +7 \\ \hline \end{array}$$

| 10 | | 11 | | 9 |

5.
$$\begin{array}{r} 4 \\ +7 \\ \hline \end{array}$$

| 12 | | 13 | | 11 |

6.
$$\begin{array}{r} 3 \\ +7 \\ \hline \end{array}$$

| 10 | | 11 | | 9 |

7.
$$\begin{array}{r} 6 \\ +7 \\ \hline \end{array}$$

| 12 | | 13 | | 14 |

8.
$$\begin{array}{r} 8 \\ +7 \\ \hline \end{array}$$

| 14 | | 16 | | 15 |

Name: _____

Add the numbers and then pick the box with the correct answer to each problem.

1.
$$\begin{array}{r} 9 \\ +8 \\ \hline \end{array}$$

| 18 | 15 | 17 |

2.
$$\begin{array}{r} 5 \\ +8 \\ \hline \end{array}$$

| 12 | 14 | 13 |

3.
$$\begin{array}{r} 7 \\ +8 \\ \hline \end{array}$$

| 16 | 15 | 14 |

4.
$$\begin{array}{r} 2 \\ +8 \\ \hline \end{array}$$

| 11 | 12 | 10 |

5.
$$\begin{array}{r} 4 \\ +8 \\ \hline \end{array}$$

| 13 | 14 | 12 |

6.
$$\begin{array}{r} 3 \\ +8 \\ \hline \end{array}$$

| 11 | 12 | 10 |

7.
$$\begin{array}{r} 6 \\ +8 \\ \hline \end{array}$$

| 13 | 14 | 15 |

8.
$$\begin{array}{r} 8 \\ +8 \\ \hline \end{array}$$

| 15 | 17 | 16 |

Name: _____

Add the numbers and then pick the box with the correct answer to each problem.

1.
$$\begin{array}{r} 9 \\ +9 \\ \hline \end{array}$$

| 19 | 16 | 18 |

2.
$$\begin{array}{r} 5 \\ +9 \\ \hline \end{array}$$

| 13 | 15 | 14 |

3.
$$\begin{array}{r} 7 \\ +9 \\ \hline \end{array}$$

| 17 | 16 | 15 |

4.
$$\begin{array}{r} 2 \\ +9 \\ \hline \end{array}$$

| 12 | 13 | 11 |

5.
$$\begin{array}{r} 4 \\ +9 \\ \hline \end{array}$$

| 14 | 15 | 13 |

6.
$$\begin{array}{r} 3 \\ +9 \\ \hline \end{array}$$

| 12 | 13 | 11 |

7.
$$\begin{array}{r} 6 \\ +9 \\ \hline \end{array}$$

| 14 | 15 | 16 |

8.
$$\begin{array}{r} 8 \\ +9 \\ \hline \end{array}$$

| 16 | 18 | 17 |

Unit 6 - Finding Addition Patterns

Now that you know your basic math facts, you can use this knowledge to solve other addition problems.

In order to use your knowledge of math facts to help you solve bigger problems you need to look for the pattern in the problem.

You know that 9 + 3 = 12.

9 + 13 is asking you to add 10 more to the problem. That is the pattern. If 9 + 3 = 12 then 9 + 3 = 12 + 10 = 23.

9	9	9	9	9	9	9
+ 3	+ 13	+ 23	+ 33	+ 43	+ 53	+ 63
12	22	32	42	52	62	72

Follow the pattern to solve each problem.

1.

10	10	10	10	10	10
+ 2	+ 12	+ 22	+ 32	+ 42	+ 52

2.

2	12	22	32	42	52
+ 5	+ 5	+ 5	+ 5	+ 5	+ 5

3.

8	18	28	38	48	58
+ 3	+ 3	+ 3	+ 3	+ 3	+ 3

Name: _____

Follow the pattern to solve each problem.

1. $\begin{array}{r} 7 \\ +\,5 \\ \hline \end{array}$ $\begin{array}{r} 17 \\ +\,5 \\ \hline \end{array}$ $\begin{array}{r} 27 \\ +\,5 \\ \hline \end{array}$ $\begin{array}{r} 37 \\ +\,5 \\ \hline \end{array}$ $\begin{array}{r} 47 \\ +\,5 \\ \hline \end{array}$ $\begin{array}{r} 57 \\ +\,5 \\ \hline \end{array}$

2. $\begin{array}{r} 6 \\ +\,8 \\ \hline \end{array}$ $\begin{array}{r} 6 \\ +\,18 \\ \hline \end{array}$ $\begin{array}{r} 6 \\ +\,28 \\ \hline \end{array}$ $\begin{array}{r} 6 \\ +\,38 \\ \hline \end{array}$ $\begin{array}{r} 6 \\ +\,48 \\ \hline \end{array}$ $\begin{array}{r} 6 \\ +\,58 \\ \hline \end{array}$

3. $\begin{array}{r} 5 \\ +\,9 \\ \hline \end{array}$ $\begin{array}{r} 15 \\ +\,9 \\ \hline \end{array}$ $\begin{array}{r} 25 \\ +\,9 \\ \hline \end{array}$ $\begin{array}{r} 35 \\ +\,9 \\ \hline \end{array}$ $\begin{array}{r} 45 \\ +\,9 \\ \hline \end{array}$ $\begin{array}{r} 55 \\ +\,9 \\ \hline \end{array}$

4. $\begin{array}{r} 3 \\ +\,7 \\ \hline \end{array}$ $\begin{array}{r} 3 \\ +\,17 \\ \hline \end{array}$ $\begin{array}{r} 3 \\ +\,27 \\ \hline \end{array}$ $\begin{array}{r} 3 \\ +\,37 \\ \hline \end{array}$ $\begin{array}{r} 3 \\ +\,47 \\ \hline \end{array}$ $\begin{array}{r} 3 \\ +\,57 \\ \hline \end{array}$

5. $\begin{array}{r} 4 \\ +\,8 \\ \hline \end{array}$ $\begin{array}{r} 14 \\ +\,8 \\ \hline \end{array}$ $\begin{array}{r} 24 \\ +\,8 \\ \hline \end{array}$ $\begin{array}{r} 34 \\ +\,8 \\ \hline \end{array}$ $\begin{array}{r} 44 \\ +\,8 \\ \hline \end{array}$ $\begin{array}{r} 54 \\ +\,8 \\ \hline \end{array}$

6. $\begin{array}{r} 9 \\ +\,6 \\ \hline \end{array}$ $\begin{array}{r} 19 \\ +\,6 \\ \hline \end{array}$ $\begin{array}{r} 29 \\ +\,6 \\ \hline \end{array}$ $\begin{array}{r} 39 \\ +\,6 \\ \hline \end{array}$ $\begin{array}{r} 49 \\ +\,6 \\ \hline \end{array}$ $\begin{array}{r} 59 \\ +\,6 \\ \hline \end{array}$

Name: _____

Follow the pattern to solve each problem.

1.
$$\begin{array}{r} 4 \\ +\ 7 \\ \hline \end{array}$$
$$\begin{array}{r} 4 \\ +\ 17 \\ \hline \end{array}$$
$$\begin{array}{r} 4 \\ +\ 27 \\ \hline \end{array}$$
$$\begin{array}{r} 4 \\ +\ 37 \\ \hline \end{array}$$
$$\begin{array}{r} 4 \\ +\ 47 \\ \hline \end{array}$$
$$\begin{array}{r} 4 \\ +\ 57 \\ \hline \end{array}$$

2.
$$\begin{array}{r} 9 \\ +\ 8 \\ \hline \end{array}$$
$$\begin{array}{r} 19 \\ +\ 8 \\ \hline \end{array}$$
$$\begin{array}{r} 29 \\ +\ 8 \\ \hline \end{array}$$
$$\begin{array}{r} 39 \\ +\ 8 \\ \hline \end{array}$$
$$\begin{array}{r} 49 \\ +\ 8 \\ \hline \end{array}$$
$$\begin{array}{r} 59 \\ +\ 8 \\ \hline \end{array}$$

3.
$$\begin{array}{r} 6 \\ +\ 6 \\ \hline \end{array}$$
$$\begin{array}{r} 6 \\ +\ 16 \\ \hline \end{array}$$
$$\begin{array}{r} 6 \\ +\ 26 \\ \hline \end{array}$$
$$\begin{array}{r} 6 \\ +\ 36 \\ \hline \end{array}$$
$$\begin{array}{r} 6 \\ +\ 46 \\ \hline \end{array}$$
$$\begin{array}{r} 6 \\ +\ 56 \\ \hline \end{array}$$

4.
$$\begin{array}{r} 5 \\ +\ 7 \\ \hline \end{array}$$
$$\begin{array}{r} 15 \\ +\ 7 \\ \hline \end{array}$$
$$\begin{array}{r} 25 \\ +\ 7 \\ \hline \end{array}$$
$$\begin{array}{r} 35 \\ +\ 7 \\ \hline \end{array}$$
$$\begin{array}{r} 45 \\ +\ 7 \\ \hline \end{array}$$
$$\begin{array}{r} 55 \\ +\ 7 \\ \hline \end{array}$$

5.
$$\begin{array}{r} 8 \\ +\ 8 \\ \hline \end{array}$$
$$\begin{array}{r} 8 \\ +\ 18 \\ \hline \end{array}$$
$$\begin{array}{r} 8 \\ +\ 28 \\ \hline \end{array}$$
$$\begin{array}{r} 8 \\ +\ 38 \\ \hline \end{array}$$
$$\begin{array}{r} 8 \\ +\ 48 \\ \hline \end{array}$$
$$\begin{array}{r} 8 \\ +\ 58 \\ \hline \end{array}$$

6.
$$\begin{array}{r} 9 \\ +\ 9 \\ \hline \end{array}$$
$$\begin{array}{r} 19 \\ +\ 9 \\ \hline \end{array}$$
$$\begin{array}{r} 29 \\ +\ 9 \\ \hline \end{array}$$
$$\begin{array}{r} 39 \\ +\ 9 \\ \hline \end{array}$$
$$\begin{array}{r} 49 \\ +\ 9 \\ \hline \end{array}$$
$$\begin{array}{r} 59 \\ +\ 9 \\ \hline \end{array}$$

Unit 7 - Number Lines and Bigger Numbers

Instructor Note:

Your student should gain the ability to use a number line in place of concrete objects to help solve bigger number addition problems.

Student Instruction:

Number lines can help you add bigger numbers too. The number line will have a beginning number and an ending number.

This number line begins at the number 0 and ends at the number 24. That means that you can add any numbers whose answer is 0 and higher or 24 and lower. You can make a number line that starts and ends at any number.

Use this number line to add 12 + 11

Place your pencil in the number 12.

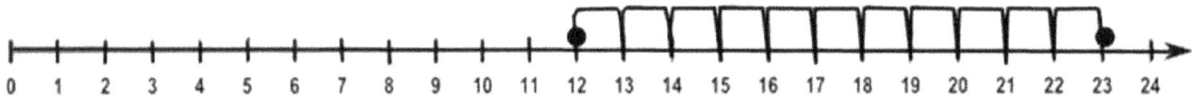

Jump your pencil forward 11 places landing on 13, 14, 15, 16, 17, 18, 19, 20, 21, 22, and 23. The last number you land on is the answer.

12 + 11 = 23

Use the number line to solve each problem.

1.

Use the number line below to solve 14 + 9 = _____

2.

Use the number line below to solve 15 + 4 = _____

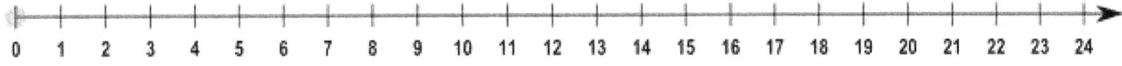

3.

Use the number line below to solve 9 + 11 = _____

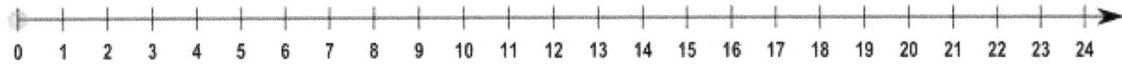

4.

Use the number line below to solve 11 + 12 = _____

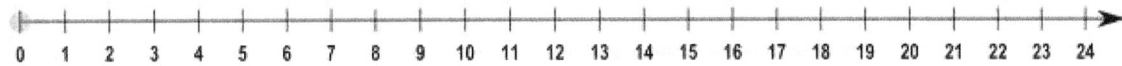

5.

Use the number line below to solve 13 + 8 = _____

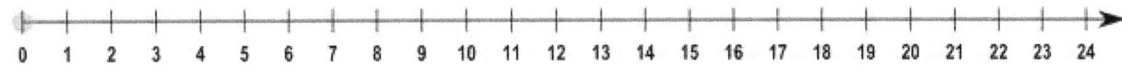

Name: _____

Use the number line to solve each problem.

1.

Use the number line below to solve 10 + 9 = _____

2.

Use the number line below to solve 16 + 8 = _____

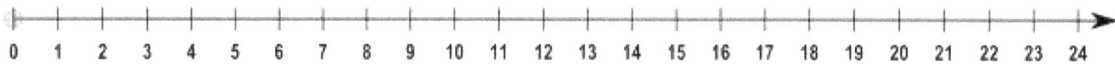

3.

Use the number line below to solve 12 + 5 = _____

4.

Use the number line below to solve 8 + 14 = _____

Name: _____

Use the number line to solve each problem.

1.

    Use the number line below to solve 9 + 13 = _____

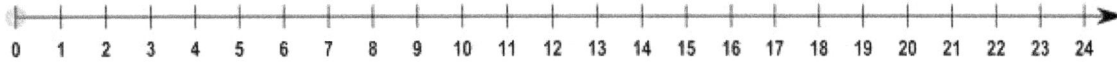

2.

    Use the number line below to solve 17 + 4 = _____

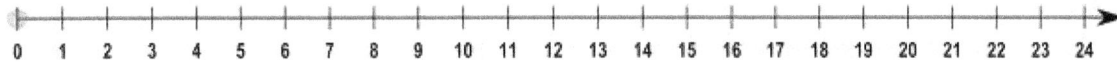

3.

    Use the number line below to solve 6 + 12 = _____

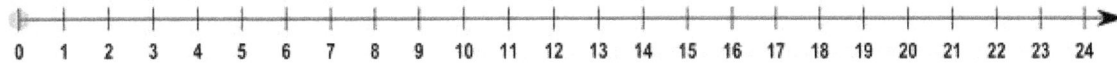

4.

    Use the number line below to solve 22 + 2 = _____

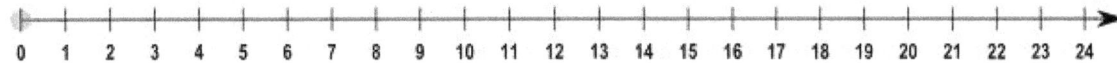

Name: _____

Use the number line to solve each problem.

1.

Use the number line below to solve 18 + 4 = _____

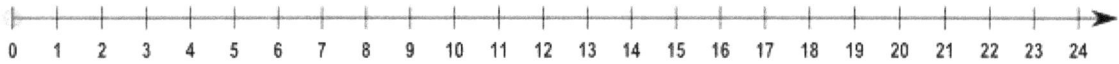

```
├──┼──┼──┼──┼──┼──┼──┼──┼──┼──┼──┼──┼──┼──┼──┼──┼──┼──┼──┼──┼──┼──┼──┼──►
0  1  2  3  4  5  6  7  8  9  10 11 12 13 14 15 16 17 18 19 20 21 22 23 24
```

2.

Use the number line below to solve 7 + 11 = _____

```
├──┼──┼──┼──┼──┼──┼──┼──┼──┼──┼──┼──┼──┼──┼──┼──┼──┼──┼──┼──┼──┼──┼──┼──►
0  1  2  3  4  5  6  7  8  9  10 11 12 13 14 15 16 17 18 19 20 21 22 23 24
```

3.

Use the number line below to solve 19 + 5 = _____

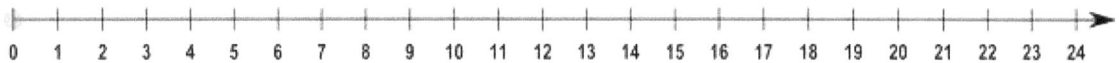

```
├──┼──┼──┼──┼──┼──┼──┼──┼──┼──┼──┼──┼──┼──┼──┼──┼──┼──┼──┼──┼──┼──┼──┼──►
0  1  2  3  4  5  6  7  8  9  10 11 12 13 14 15 16 17 18 19 20 21 22 23 24
```

4.

Use the number line below to solve 5 + 15 = _____

```
├──┼──┼──┼──┼──┼──┼──┼──┼──┼──┼──┼──┼──┼──┼──┼──┼──┼──┼──┼──┼──┼──┼──┼──►
0  1  2  3  4  5  6  7  8  9  10 11 12 13 14 15 16 17 18 19 20 21 22 23 24
```

Name: _____

Use the number line to solve each problem.

1.

Use the number line below to solve 20 + 2 = _____

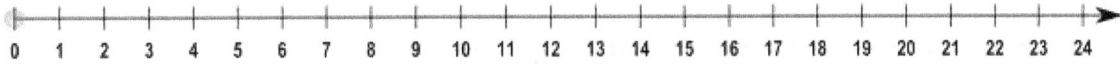

2.

Use the number line below to solve 21 + 3 = _____

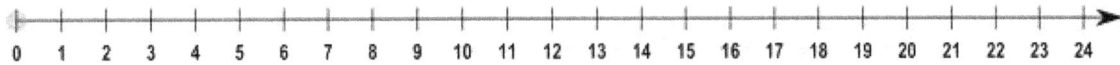

3.

Use the number line below to solve 8 + 11 = _____

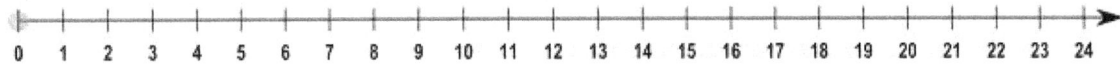

4.

Use the number line below to solve 3 + 20 = _____

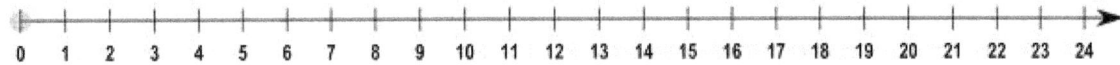

Unit 8 - Addition Using a 100's Chart

Instructor Note:

Your student should transition to higher number addition with the aid of a hundred chart.

Student Instruction:

A 100's chart is another tool you can use to help you add larger numbers. The 100's chart begins at the number 1 and ends at the number 100. That means that you can use the hundred chart to help you add any numbers whose answer is 1 or more and 100 or less.

Use the 100's chart to add 50 + 12

1	2	3	4	5	6	7	8	9	10
11	12	13	14	15	16	17	18	19	20
21	22	23	24	25	26	27	28	29	30
31	32	33	34	35	36	37	38	39	40
41	42	43	44	45	46	47	48	49	50
51	52	53	54	55	56	57	58	59	60
61	62	63	64	65	66	67	68	69	70
71	72	73	74	75	76	77	78	79	80
81	82	83	84	85	86	87	88	89	90
91	92	93	94	95	96	97	98	99	100

Place your pencil on the number 50.

Jump your pencil forward 12 places landing on 51, 52, 53, 54, 55, 56, 57, 58, 59, 60, 61, and 62. The last number you land on is the answer.

50 + 12 = 62

1	2	3	4	5	6	7	8	9	10
11	12	13	14	15	16	17	18	19	20
21	22	23	24	25	26	27	28	29	30
31	32	33	34	35	36	37	38	39	40
41	42	43	44	45	46	47	48	49	50
51	52	53	54	55	56	57	58	59	60
61	62	63	64	65	66	67	68	69	70
71	72	73	74	75	76	77	78	79	80
81	82	83	84	85	86	87	88	89	90
91	92	93	94	95	96	97	98	99	100

Cut out the 100's chart and use it to solve each problem.

Name: _____

Use the 100 chart to help you solve each problem.

1.
$$\begin{array}{r} 70 \\ +6 \\ \hline \end{array}$$

2.
$$\begin{array}{r} 55 \\ +2 \\ \hline \end{array}$$

3.
$$\begin{array}{r} 18 \\ +5 \\ \hline \end{array}$$

4.
$$\begin{array}{r} 22 \\ +3 \\ \hline \end{array}$$

5.
$$\begin{array}{r} 28 \\ +9 \\ \hline \end{array}$$

6.
$$\begin{array}{r} 38 \\ +4 \\ \hline \end{array}$$

7.
$$\begin{array}{r} 31 \\ +7 \\ \hline \end{array}$$

8.
$$\begin{array}{r} 61 \\ +8 \\ \hline \end{array}$$

9.
$$\begin{array}{r} 25 \\ +9 \\ \hline \end{array}$$

10.
$$\begin{array}{r} 42 \\ +8 \\ \hline \end{array}$$

Name: _____

Use the 100 chart to help you solve each problem.

1.
$$\begin{array}{r} 61 \\ +5 \\ \hline \end{array}$$

2.
$$\begin{array}{r} 46 \\ +2 \\ \hline \end{array}$$

3.
$$\begin{array}{r} 29 \\ +5 \\ \hline \end{array}$$

4.
$$\begin{array}{r} 33 \\ +3 \\ \hline \end{array}$$

5.
$$\begin{array}{r} 17 \\ +9 \\ \hline \end{array}$$

6.
$$\begin{array}{r} 56 \\ +4 \\ \hline \end{array}$$

7.
$$\begin{array}{r} 62 \\ +7 \\ \hline \end{array}$$

8.
$$\begin{array}{r} 74 \\ +8 \\ \hline \end{array}$$

9.
$$\begin{array}{r} 95 \\ +3 \\ \hline \end{array}$$

10.
$$\begin{array}{r} 87 \\ +8 \\ \hline \end{array}$$

Name: _____

Use the 100 chart to help you solve each problem.

1.
$$59$$
$$+21$$

2.
$$32$$
$$+30$$

3.
$$87$$
$$+11$$

4.
$$64$$
$$+19$$

5.
$$21$$
$$+16$$

6.
$$19$$
$$+11$$

7.
$$46$$
$$+12$$

8.
$$78$$
$$+14$$

9.
$$90$$
$$+8$$

10.
$$53$$
$$+18$$

Name: _____

Use the 100 chart to help you solve each problem.

1.
$$70 \atop +25$$

2.
$$55 \atop +32$$

3.
$$18 \atop +27$$

4.
$$22 \atop +34$$

5.
$$28 \atop +29$$

6.
$$38 \atop +26$$

7.
$$31 \atop +23$$

8.
$$61 \atop +18$$

9.
$$25 \atop +39$$

10.
$$42 \atop +38$$

Name: _____

Use the 100 chart to help you solve each problem.

1.
$$67$$
$$+25$$

2.
$$42$$
$$+32$$

3.
$$38$$
$$+27$$

4.
$$26$$
$$+34$$

5.
$$54$$
$$+29$$

6.
$$71$$
$$+26$$

7.
$$70$$
$$+23$$

8.
$$64$$
$$+18$$

9.
$$25$$
$$+25$$

10.
$$51$$
$$+38$$

Unit 9 - 2 Digit Addition

Some problems will have more than 1 digit on each line. You can use the hundred chart or a number line to help you solve each problem. You can also do all of these problems using the math facts you have learned!

$$\begin{array}{r} 12 \\ + 15 \\ \hline \end{array}$$

You will solve these problems starting at the ones place and moving to the left.

$$\begin{array}{r} 12 \\ + 15 \\ \hline 7 \end{array} \qquad \begin{array}{r} 12 \\ + 15 \\ \hline 27 \end{array}$$

Solve each problem using your math facts.

1. $\begin{array}{r} 56 \\ + 11 \\ \hline \end{array}$ 2. $\begin{array}{r} 40 \\ + 16 \\ \hline \end{array}$ 3. $\begin{array}{r} 36 \\ + 12 \\ \hline \end{array}$

4. $\begin{array}{r} 26 \\ + 32 \\ \hline \end{array}$ 5. $\begin{array}{r} 45 \\ + 14 \\ \hline \end{array}$ 6. $\begin{array}{r} 32 \\ + 66 \\ \hline \end{array}$

7. $\begin{array}{r} 65 \\ + 22 \\ \hline \end{array}$ 8. $\begin{array}{r} 71 \\ + 27 \\ \hline \end{array}$ 9. $\begin{array}{r} 83 \\ + 15 \\ \hline \end{array}$

Name: _____

Solve each problem using your math facts.

1.   56
      + 11

2.   40
      + 16

3.   36
      + 12

4.   26
      + 32

5.   45
      + 14

6.   32
      + 66

7.   65
      + 22

8.   71
      + 27

9.   83
      + 15

10.   51
      + 36

11.   51
      + 37

12.   67
      + 12

Name: _____

Solve each problem using your math facts.

1. $\begin{array}{r} 45 \\ + 21 \end{array}$ 2. $\begin{array}{r} 39 \\ + 40 \end{array}$ 3. $\begin{array}{r} 25 \\ + 61 \end{array}$

4. $\begin{array}{r} 16 \\ + 21 \end{array}$ 5. $\begin{array}{r} 34 \\ + 53 \end{array}$ 6. $\begin{array}{r} 21 \\ + 55 \end{array}$

7. $\begin{array}{r} 54 \\ + 11 \end{array}$ 8. $\begin{array}{r} 60 \\ + 16 \end{array}$ 9. $\begin{array}{r} 72 \\ + 24 \end{array}$

10. $\begin{array}{r} 73 \\ + 26 \end{array}$ 11. $\begin{array}{r} 59 \\ + 40 \end{array}$ 12. $\begin{array}{r} 82 \\ + 16 \end{array}$

Name: _____

Solve each problem using your math facts.

1.   34
    + 32

2.   27
    + 62

3.   36
    + 52

4.   27
    + 32

5.   45
    + 42

6.   32
    + 66

7.   65
    + 22

8.   71
    + 26

9.   83
    + 13

10.   57
    + 32

11.   70
    + 29

12.   11
    + 87

Name: _____

Solve each problem using your math facts.

1. $\begin{array}{r} 45 \\ +33 \\ \hline \end{array}$	2. $\begin{array}{r} 38 \\ +51 \\ \hline \end{array}$	3. $\begin{array}{r} 47 \\ +52 \\ \hline \end{array}$
4. $\begin{array}{r} 38 \\ +41 \\ \hline \end{array}$	5. $\begin{array}{r} 56 \\ +31 \\ \hline \end{array}$	6. $\begin{array}{r} 43 \\ +55 \\ \hline \end{array}$
7. $\begin{array}{r} 76 \\ +11 \\ \hline \end{array}$	8. $\begin{array}{r} 82 \\ +16 \\ \hline \end{array}$	9. $\begin{array}{r} 25 \\ +24 \\ \hline \end{array}$
10. $\begin{array}{r} 68 \\ +21 \\ \hline \end{array}$	11. $\begin{array}{r} 60 \\ +19 \\ \hline \end{array}$	12. $\begin{array}{r} 22 \\ +76 \\ \hline \end{array}$

Unit 10 - 3 Digit Addition

Student Instruction:

Some problems will have more than 1 digit on each line. You can do all of these problems using the math facts you have learned!

$$
\begin{array}{r}
537 \\
+\,322 \\
\hline
\end{array}
$$

You will solve these problems starting at the ones place and moving to the left.

$$
\begin{array}{r}
537 \\
+\,322 \\
\hline
9
\end{array}
\qquad
\begin{array}{r}
537 \\
+\,322 \\
\hline
59
\end{array}
\qquad
\begin{array}{r}
537 \\
+\,322 \\
\hline
859
\end{array}
$$

Solve each problem using your math facts.

1.
$$
\begin{array}{r}
304 \\
+\,152 \\
\hline
\end{array}
$$

2.
$$
\begin{array}{r}
236 \\
+\,211 \\
\hline
\end{array}
$$

3.
$$
\begin{array}{r}
356 \\
+\,142 \\
\hline
\end{array}
$$

4.
$$
\begin{array}{r}
526 \\
+\,432 \\
\hline
\end{array}
$$

5.
$$
\begin{array}{r}
445 \\
+\,214 \\
\hline
\end{array}
$$

6.
$$
\begin{array}{r}
632 \\
+\,366 \\
\hline
\end{array}
$$

7.
$$
\begin{array}{r}
465 \\
+\,122 \\
\hline
\end{array}
$$

8.
$$
\begin{array}{r}
771 \\
+\,227 \\
\hline
\end{array}
$$

9.
$$
\begin{array}{r}
883 \\
+\,115 \\
\hline
\end{array}
$$

Name: _____

Solve each problem using your math facts.

1.	526 + 131	2.	460 + 126	3.	356 + 142

4.	726 + 232	5.	145 + 814	6.	302 + 696

7.	265 + 522	8.	701 + 297	9.	383 + 415

10.	685 + 212	11.	271 + 527	12.	853 + 145

Name: _____

Solve each problem using your math facts.

1.   345
    + 221
	2. 639
	    + 240
		3. 295
		    + 601

4.   816
    + 121
	5. 374
	    + 523
		6. 721
		    + 155

7.   654
    + 311
	8. 670
	    + 126
		9. 742
		    + 254

10.   573
    + 426
	11. 659
	    + 340
		12. 482
		    + 416

Name: _____

Solve each problem using your math facts.

1.	324 + 372	2.	207 + 692	3.	436 + 352

4.	267 + 332	5.	495 + 402	6.	232 + 766

7.	765 + 122	8.	741 + 256	9.	383 + 613

10.	557 + 432	11.	870 + 129	12.	161 + 837

Name: _____

Solve each problem using your math facts.

1. 495
    + 303

2. 738
    + 251

3. 547
    + 352

4. 378
    + 421

5. 656
    + 231

6. 343
    + 455

7. 876
    + 121

8. 382
    + 516

9. 625
    + 224

10. 368
    + 521

11. 670
    + 129

12. 232
    + 756

Unit 11 - Lining Up Numbers for Addition

Instructor Note:

Your student should understand how to line up horizontal addition problems in a vertical format.

Student Instruction:

Sometimes a math problem will appear in a line.

31 + 17 = 48

Other times a math problem will appear in a stack.

```
      3    1
 +    1    7
_____

      4    8
```

You should always rearrange the problem so that it lines up in a stack. This will help you to solve larger number math problems.

Write the first number in the math problem on your paper.

```
      3    1
```

Write the second number underneath so that the ones place of each number lines up with the one above it.

```
      3    1
      1    7
```

Add the symbol that tells you this is an addition problem.

```
      3    1
 +    1    7
```

Draw a line under the bottom number to show where the answer to the problem begins.

```
      3    1
 +    1    7
_____
```

Now the problem is lined up so that you know which numbers to add together.

You will add the numbers starting with the smallest place value and adding one column at a time moving left to the largest place value.

You will begin by adding the ones column.

```
    3   1
+   1   7
_____
        8
```

Next, you will add the numbers in the tens column.

```
    3   1
+   1   7
_____
    4   8
```

If the numbers were larger, you would continue working right to left until you have completed the calculations for each column.

Name: _____

Line up each problem and then solve it.

1.

    12 + 5 = _____

_____

2.

    11 + 4 = _____

_____

3.

    11 + 8 = _____

_____

4.

    12 + 7 = _____

_____

5.

    13 + 5 = _____

_____

Name: _____

Line up each problem and then solve it.

1.

  10 + 29 = _____

_____

2.

  16 + 22 = _____

_____

3.

  12 + 15 = _____

_____

4.

  17 + 12 = _____

_____

5.

  18 + 11 = _____

_____

Name: _____

Line up each problem and then solve it.

1.

   624 + 35 = _____

_____

2.

   275 + 24 = _____

_____

3.

   451 + 48 = _____

_____

4.

   732 + 57 = _____

_____

5.

   483 + 15 = _____

_____

Line up each problem and then solve it.

1.
$540 + 219 =$ _____

2.
$386 + 612 =$ _____

3.
$862 + 125 =$ _____

4.
$327 + 432 =$ _____

5.
$738 + 221 =$ _____

Name: _____

Line up each problem and then solve it.

1.

439 + 110 = _____

_____

2.

275 + 511 = _____

_____

3.

751 + 34 = _____

_____

4.

216 + 321 = _____

_____

5.

627 + 110 = _____

_____

Unit 12 - Adding Larger Numbers

Instructor Note:

Your student should transition from adding one and two digit numbers to adding larger numbers.

Student Instruction:

Adding larger numbers is the same as adding smaller numbers. You will always begin at the right with the unit column and move to the left adding the tens, hundreds, thousands, ten thousands, hundred thousands, millions, and ten millions.

Remember your place value chart so that you know what the value of each column is before you add.

61,361,986									
Ten Millions	Millions	,	Hundred Thousands	Ten Thousands	Thousands	,	Hundreds	Tens	Ones
Six Ten Millions	One Million	,	Three Hundred Thousands	Six Ten Thousands	One Thousand	,	Nine Hundreds	Eight Tens	Six Ones
60,000,000	1,000,000	,	300,000	60,000	1,000	,	900	80	3

Remember to add the commas in the correct location to help to make your answers easier to read.   A comma goes after every 3 places moving from right to left.

Name: _____

1.
$$452$$
$$+\ 25$$

2.
$$219$$
$$+\ 30$$

3.
$$138$$
$$+\ 21$$

4.
$$526$$
$$+\ 32$$

5.
$$654$$
$$+\ 23$$

6.
$$977$$
$$+\ 12$$

7.
$$845$$
$$+\ 23$$

8.
$$664$$
$$+\ 35$$

9.
$$721$$
$$+\ 26$$

10.
$$253$$
$$+\ 34$$

Name: _____

1.
$$351 + 36$$

2.
$$118 + 41$$

3.
$$137 + 32$$

4.
$$425 + 43$$

5.
$$553 + 34$$

6.
$$876 + 23$$

7.
$$744 + 34$$

8.
$$563 + 26$$

9.
$$620 + 37$$

10.
$$252 + 45$$

Name: _____

1.
$$450$$
$$+\ 536$$

2.
$$217$$
$$+\ 741$$

3.
$$238$$
$$+\ 630$$

4.
$$524$$
$$+\ 443$$

5.
$$653$$
$$+\ 334$$

6.
$$775$$
$$+\ 223$$

7.
$$844$$
$$+\ 134$$

8.
$$163$$
$$+\ 826$$

9.
$$729$$
$$+\ 260$$

10.
$$353$$
$$+\ 645$$

Name: _____

1.
$$351$$
$$+\ 437$$

2.
$$218$$
$$+\ 641$$

3.
$$139$$
$$+\ 530$$

4.
$$423$$
$$+\ 542$$

5.
$$754$$
$$+\ 235$$

6.
$$674$$
$$+124$$

7.
$$543$$
$$+\ 353$$

8.
$$263$$
$$+\ 625$$

9.
$$628$$
$$+\ 261$$

10.
$$452$$
$$+\ 546$$

Name: _____

1.
$$7,452 \\ +25$$

2.
$$4,219 \\ +30$$

3.
$$5,138 \\ +21$$

4.
$$1,526 \\ +32$$

5.
$$9,654 \\ +23$$

6.
$$3,977 \\ +12$$

7.
$$2,845 \\ +23$$

8.
$$8,664 \\ +35$$

9.
$$6,721 \\ +26$$

10.
$$7,253 \\ +34$$

Name: _____

1.
$$\begin{array}{r} 5{,}351 \\ +36 \\ \hline \end{array}$$

2.
$$\begin{array}{r} 9{,}118 \\ +41 \\ \hline \end{array}$$

3.
$$\begin{array}{r} 3{,}137 \\ +32 \\ \hline \end{array}$$

4.
$$\begin{array}{r} 6{,}425 \\ +43 \\ \hline \end{array}$$

5.
$$\begin{array}{r} 7{,}553 \\ +34 \\ \hline \end{array}$$

6.
$$\begin{array}{r} 4{,}876 \\ +23 \\ \hline \end{array}$$

7.
$$\begin{array}{r} 1{,}744 \\ +34 \\ \hline \end{array}$$

8.
$$\begin{array}{r} 3{,}563 \\ +26 \\ \hline \end{array}$$

9.
$$\begin{array}{r} 2{,}620 \\ +37 \\ \hline \end{array}$$

10.
$$\begin{array}{r} 8{,}252 \\ +45 \\ \hline \end{array}$$

Name: _____

1.
$$4,450$$
$$+\ 536$$

2.
$$7,217$$
$$+741$$

3.
$$8,238$$
$$+\ 630$$

4.
$$1,524$$
$$+\ 443$$

5.
$$3,653$$
$$+\ 334$$

6.
$$5,775$$
$$+\ 223$$

7.
$$9,844$$
$$+\ 134$$

8.
$$2,163$$
$$+\ 826$$

9.
$$6,729$$
$$+\ 260$$

10.
$$4,353$$
$$+\ 645$$

Name: _____

1.
$$2{,}231 \\ +\ 548$$

2.
$$8{,}108 \\ +\ 751$$

3.
$$5{,}029 \\ +\ 640$$

4.
$$7{,}313 \\ +\ 653$$

5.
$$1{,}624 \\ +\ 345$$

6.
$$9{,}564 \\ +\ 234$$

7.
$$3{,}433 \\ +\ 562$$

8.
$$6{,}253 \\ +\ 736$$

9.
$$9{,}518 \\ +\ 341$$

10.
$$4{,}342 \\ +\ 657$$

Name: _____

1.
$$\begin{array}{r} 4,450 \\ +536 \\ \hline \end{array}$$

2.
$$\begin{array}{r} 7,217 \\ +741 \\ \hline \end{array}$$

3.
$$\begin{array}{r} 8,238 \\ +630 \\ \hline \end{array}$$

4.
$$\begin{array}{r} 1,524 \\ +443 \\ \hline \end{array}$$

5.
$$\begin{array}{r} 3,653 \\ +334 \\ \hline \end{array}$$

6.
$$\begin{array}{r} 5,775 \\ +223 \\ \hline \end{array}$$

7.
$$\begin{array}{r} 9,844 \\ +134 \\ \hline \end{array}$$

8.
$$\begin{array}{r} 2,163 \\ +826 \\ \hline \end{array}$$

9.
$$\begin{array}{r} 6,729 \\ +260 \\ \hline \end{array}$$

10.
$$\begin{array}{r} 4,353 \\ +645 \\ \hline \end{array}$$

1.   3,145
   + 6,542

2.   7,297
   + 2,701

3.   6,138
   + 2,640

4.   8,425
   + 1,453

5.   2,535
   + 7,342

6.   3,653
   + 5.234

7.   7,522
   + 1,462

8.   1,361
   + 8,136

9.   4,629
   + 2,340

10.   5,543
    + 4,356

Name: _____

1.
$$30{,}450$$
$$+\ 9{,}536$$

2.
$$55{,}217$$
$$+\ 4{,}741$$

3.
$$14{,}238$$
$$+\ 3{,}630$$

4.
$$81{,}524$$
$$+\ 6{,}443$$

5.
$$63{,}653$$
$$+\ 1{,}334$$

6.
$$20{,}775$$
$$+\ 9{,}223$$

7.
$$51{,}844$$
$$+\ 5{,}134$$

8.
$$42{,}163$$
$$+\ 7{,}826$$

9.
$$76{,}729$$
$$+\ 1{,}260$$

10.
$$24{,}353$$
$$+\ 2{,}645$$

Name: _____

1.
$$72,231 \\ +12,548$$

2.
$$48,108 \\ +31,751$$

3.
$$55,029 \\ +23,640$$

4.
$$17,313 \\ +71,653$$

5.
$$80,624 \\ +19,345$$

6.
$$69,564 \\ +30,234$$

7.
$$33,433 \\ +56,562$$

8.
$$46,253 \\ +43,736$$

9.
$$19,518 \\ +40,341$$

10.
$$72,342 \\ +16,657$$

Name: _____

1.
$$130,450$$
$$+ 69,536$$

2.
$$455,217$$
$$+ 24,741$$

3.
$$714,238$$
$$+ 73,630$$

4.
$$381,524$$
$$+ 16,443$$

5.
$$563,653$$
$$+ 31,334$$

6.
$$220,775$$
$$+ 69,223$$

7.
$$651,844$$
$$+ 45,134$$

8.
$$842,163$$
$$+ 57,826$$

9.
$$276,729$$
$$+ 22,260$$

10.
$$524,353$$
$$+ 32,645$$

Name: _____

1. 572,231
    +402,548

2. 148,108
    +851,751

3. 755,029
    +113,640

4. 317,313
    +641,653

5. 280,624
    +709,345

6. 669,564
    +120,234

7. 433,433
    +366,562

8. 846,253
    +153,736

9. 519,518
    +370,341

10. 272,342
    +606,657

Name: _____

1.
$$5,130,450$$
$$+ 869,536$$

2.
$$7,455,217$$
$$+ 524,741$$

3.
$$3,714,238$$
$$+ 273,630$$

4.
$$1,381,524$$
$$+ 616,443$$

5.
$$4,563,653$$
$$+ 431,334$$

6.
$$8,220,775$$
$$+ 569,223$$

7.
$$2,651,844$$
$$+ 145,134$$

8.
$$6,842,163$$
$$+ 157,826$$

9.
$$4,276,729$$
$$+ 321,260$$

10.
$$1,524,353$$
$$+ 432,645$$

Unit 13 - Addition with Regrouping

Instructor Note:

You student will need to use their knowledge of units, tens, hundreds, and thousands to gain the ability to regroup within addition problems.

Student Instruction:

You should always line up addition problems so that the place value of the numbers line up one on top of the other. 238 + 116 becomes

```
      2    3    1
+     1    1    6
_____
```

When adding numbers, you will always start with the numbers at the right. This is the unit number.

Add the numbers together.

If the sum is between 0 and 9, you write the number in the correct column.

If the sum of a column is greater than or equal to ten, you will need to carry the ten to the next column on the left. This is also called regrouping.

You always begin a math problem at the smallest unit.

```
You need to add     5
                  + 7
                  ____
                   12
```

12 = 1 ten and 2 units

You will enter the unit in the correct column of the solution line.

You need to carry the ten to the tens column before completing the problem

You will finish by adding all the tens, including the ten that you carried from the unit column.

3 tens + 1 ten + 1 ten that you carried = 5 tens.

5 tens + 2 units = 52.

```
        3   3
  +     1   7
  _____
        5   2
```

To remind ourselves that we regrouped when adding, we write a little number above the column that we regrouped.

```
        1
        3   3
  +     1   7
  _____
        5   2
```

```
        1   1
        5   8   4
  +     2   3   6
  _____
        8   2   0
```

Understanding regrouping is easier if you remember your base 10 blocks. You should always arrange the numbers into the largest block that they can make.

The unit blocks can only hold 9 before they equal a 10 bar.

Stack them and see. In one column, place a ten bar. Line up 9 units beside it. Add the 10th unit. Do you see how 10 units are equal to a 10 bar? Whenever you get 10 units, you must carry the 10 over to the tens column.

The ten bars and the hundred block work the same way.

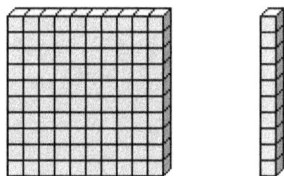

Stack them and see. Lay down a hundred block. Now, stack 9 ten bars on top of the hundred block. Add the 10th ten bar. Do you see how the ten bars completely cover the hundred block? Whenever you get 10 ten bars, you must trade it in for one hundred block. This is called regrouping.

If you have a base 10 set, use it to help you solve the problems on the following pages. If you do not have your own base 10 set, cut out the examples on these pages to help you decide when to regroup and when to add without regrouping.

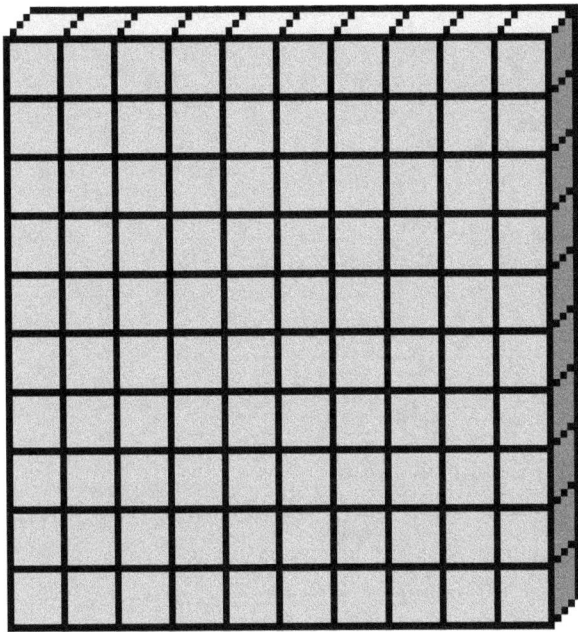
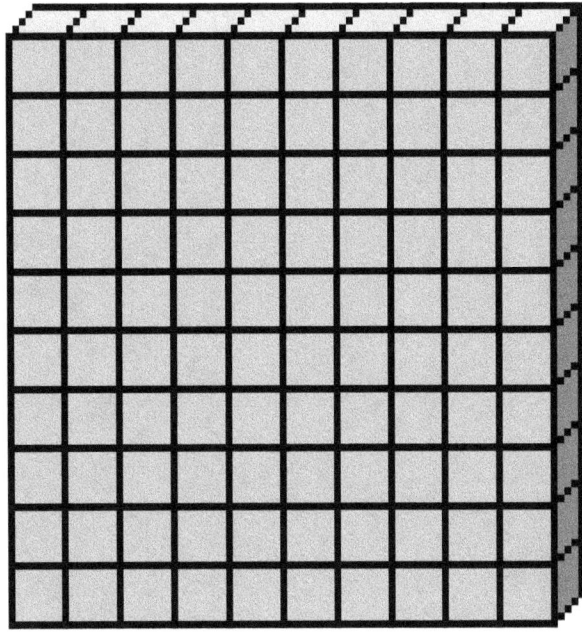
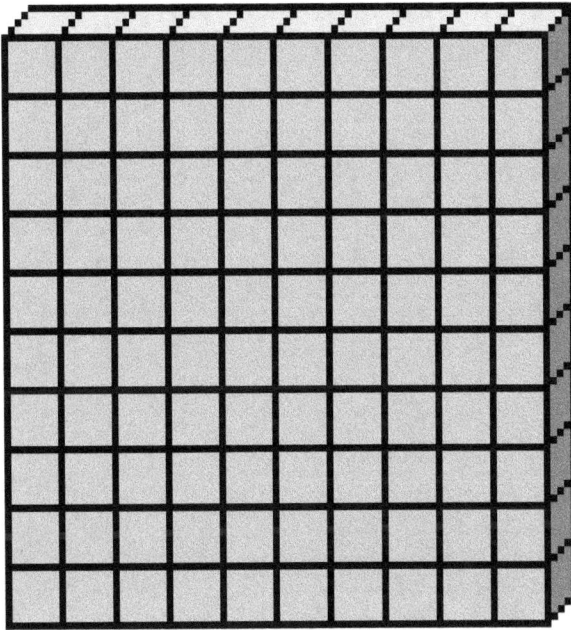
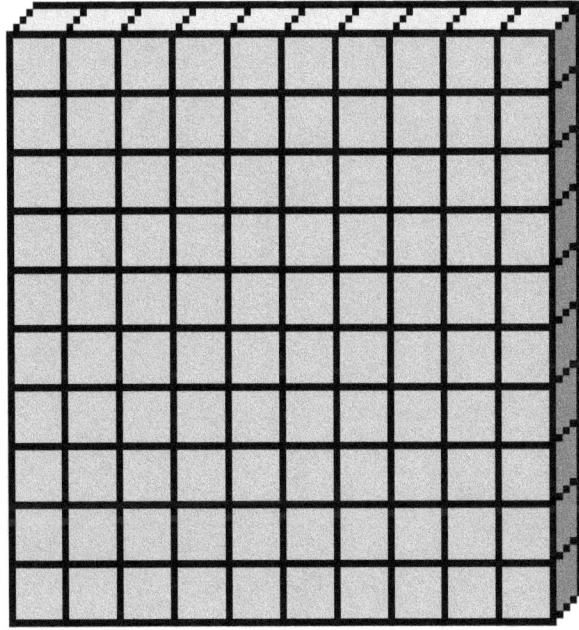

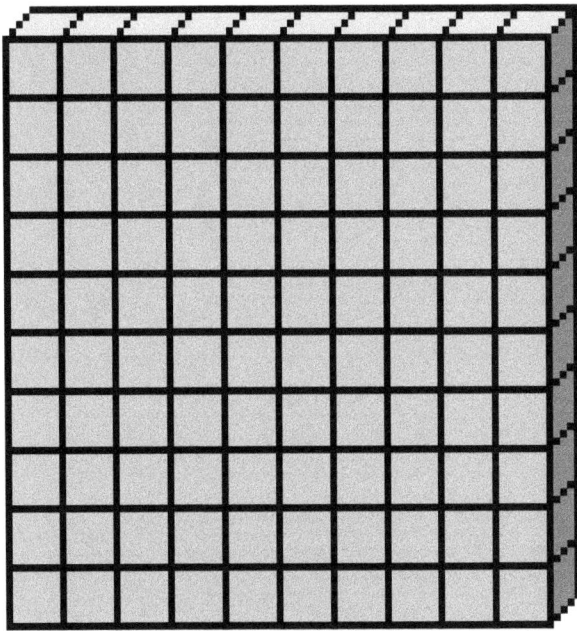
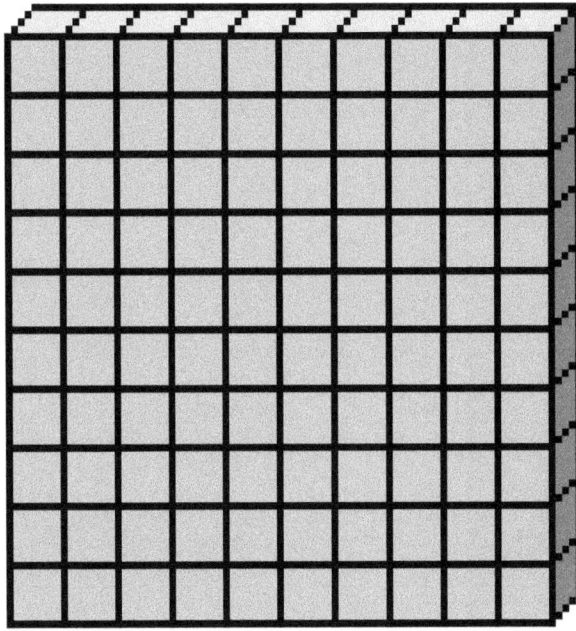
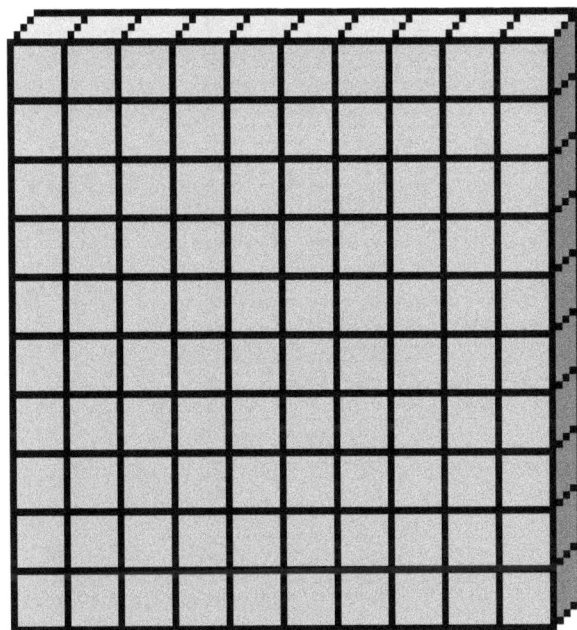
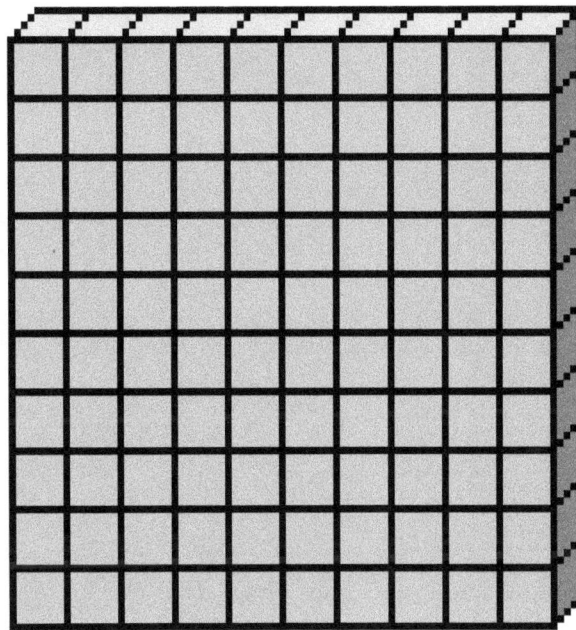
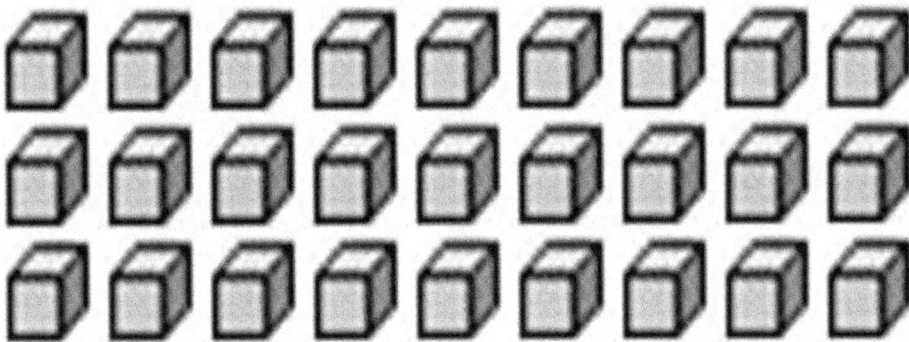

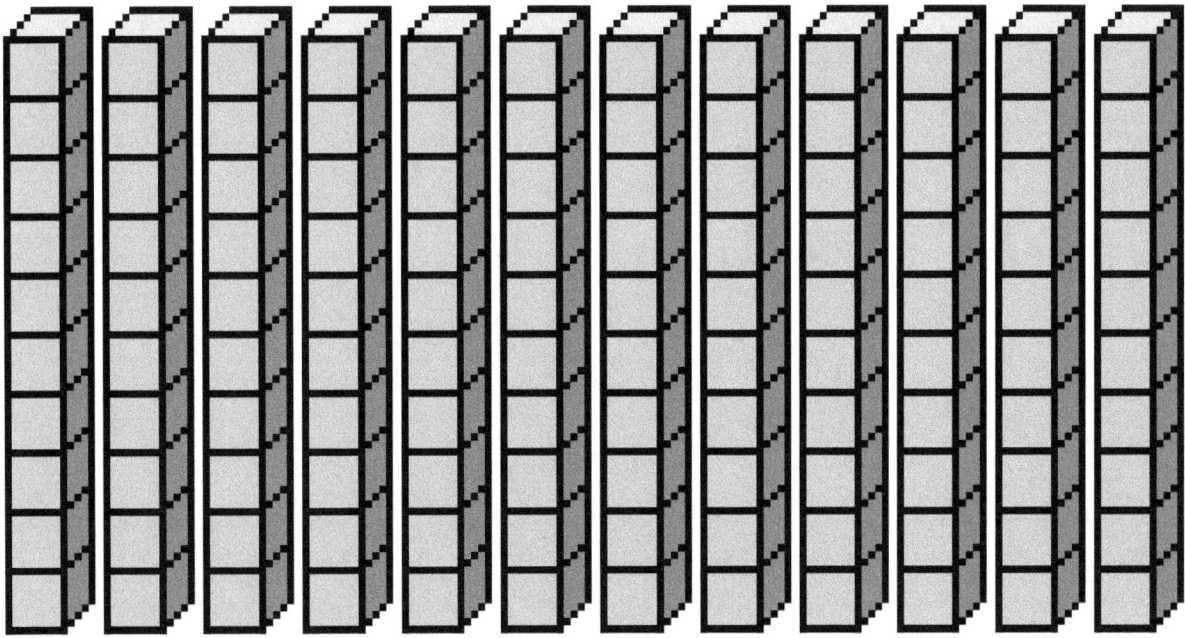

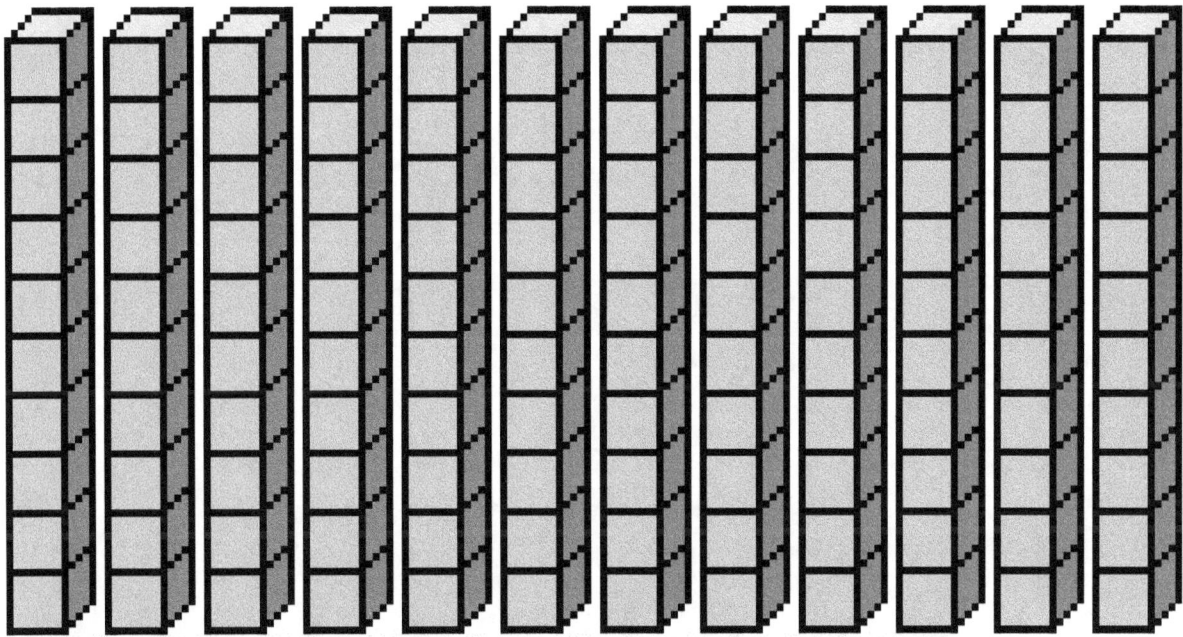

117

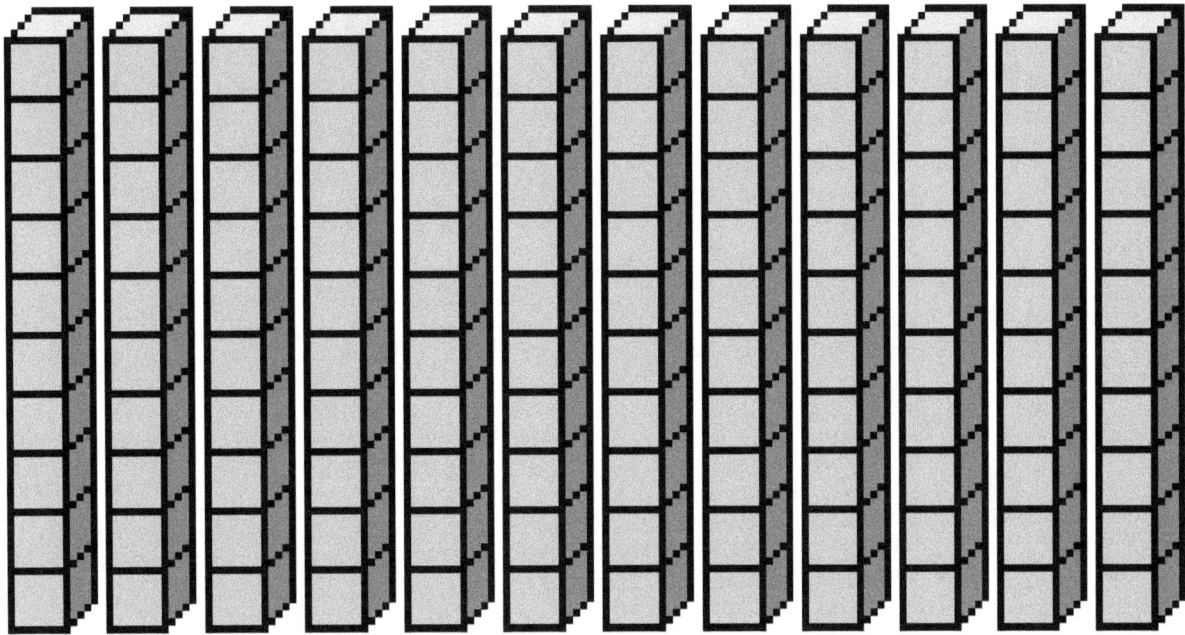

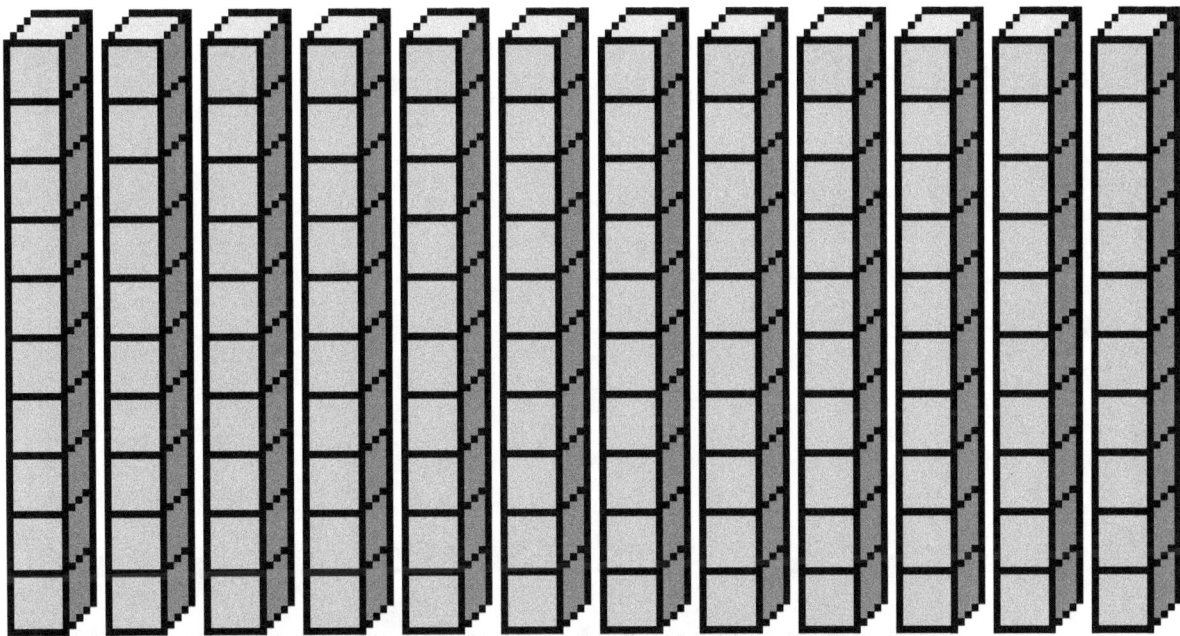

119

Name: _____

1.
$$\begin{array}{r} 67 \\ +25 \\ \hline \end{array}$$

2.
$$\begin{array}{r} 49 \\ +32 \\ \hline \end{array}$$

3.
$$\begin{array}{r} 38 \\ +27 \\ \hline \end{array}$$

4.
$$\begin{array}{r} 26 \\ +34 \\ \hline \end{array}$$

5.
$$\begin{array}{r} 54 \\ +29 \\ \hline \end{array}$$

6.
$$\begin{array}{r} 47 \\ +26 \\ \hline \end{array}$$

7.
$$\begin{array}{r} 38 \\ +23 \\ \hline \end{array}$$

8.
$$\begin{array}{r} 64 \\ +18 \\ \hline \end{array}$$

9.
$$\begin{array}{r} 29 \\ +25 \\ \hline \end{array}$$

10.
$$\begin{array}{r} 57 \\ +38 \\ \hline \end{array}$$

Name: _____

1.
$$\begin{array}{r} 18 \\ +15 \\ \hline \end{array}$$

2.
$$\begin{array}{r} 36 \\ +28 \\ \hline \end{array}$$

3.
$$\begin{array}{r} 52 \\ +29 \\ \hline \end{array}$$

4.
$$\begin{array}{r} 47 \\ +15 \\ \hline \end{array}$$

5.
$$\begin{array}{r} 76 \\ +17 \\ \hline \end{array}$$

6.
$$\begin{array}{r} 62 \\ +28 \\ \hline \end{array}$$

7.
$$\begin{array}{r} 55 \\ +27 \\ \hline \end{array}$$

8.
$$\begin{array}{r} 72 \\ +19 \\ \hline \end{array}$$

9.
$$\begin{array}{r} 39 \\ +25 \\ \hline \end{array}$$

10.
$$\begin{array}{r} 17 \\ +56 \\ \hline \end{array}$$

Name: _____

1.
$$\begin{array}{r} 24 \\ +47 \\ \hline \end{array}$$

2.
$$\begin{array}{r} 58 \\ +12 \\ \hline \end{array}$$

3.
$$\begin{array}{r} 13 \\ +39 \\ \hline \end{array}$$

4.
$$\begin{array}{r} 61 \\ +24 \\ \hline \end{array}$$

5.
$$\begin{array}{r} 37 \\ +35 \\ \hline \end{array}$$

6.
$$\begin{array}{r} 19 \\ +34 \\ \hline \end{array}$$

7.
$$\begin{array}{r} 22 \\ +49 \\ \hline \end{array}$$

8.
$$\begin{array}{r} 75 \\ +19 \\ \hline \end{array}$$

9.
$$\begin{array}{r} 32 \\ +29 \\ \hline \end{array}$$

10.
$$\begin{array}{r} 26 \\ +17 \\ \hline \end{array}$$

1.
$$452$$
$$+\ 65$$

2.
$$219$$
$$+\ 32$$

3.
$$138$$
$$+\ 71$$

4.
$$526$$
$$+\ 92$$

5.
$$654$$
$$+\ 28$$

6.
$$977$$
$$+\ 15$$

7.
$$845$$
$$+\ 83$$

8.
$$684$$
$$+\ 35$$

9.
$$721$$
$$+\ 69$$

10.
$$253$$
$$+\ 94$$

1.
$$\begin{array}{r} 313 \\ +58 \\ \hline \end{array}$$

2.
$$\begin{array}{r} 847 \\ +23 \\ \hline \end{array}$$

3.
$$\begin{array}{r} 624 \\ +58 \\ \hline \end{array}$$

4.
$$\begin{array}{r} 472 \\ +29 \\ \hline \end{array}$$

5.
$$\begin{array}{r} 148 \\ +36 \\ \hline \end{array}$$

6.
$$\begin{array}{r} 227 \\ +48 \\ \hline \end{array}$$

7.
$$\begin{array}{r} 523 \\ +59 \\ \hline \end{array}$$

8.
$$\begin{array}{r} 764 \\ +28 \\ \hline \end{array}$$

9.
$$\begin{array}{r} 543 \\ +57 \\ \hline \end{array}$$

10.
$$\begin{array}{r} 337 \\ +27 \\ \hline \end{array}$$

Name: _____

1.
$$\begin{array}{r} 227 \\ +36 \\ \hline \end{array}$$

2.
$$\begin{array}{r} 545 \\ +\ 38 \\ \hline \end{array}$$

3.
$$\begin{array}{r} 154 \\ +18 \\ \hline \end{array}$$

4.
$$\begin{array}{r} 657 \\ +\ 25 \\ \hline \end{array}$$

5.
$$\begin{array}{r} 465 \\ +26 \\ \hline \end{array}$$

6.
$$\begin{array}{r} 362 \\ +19 \\ \hline \end{array}$$

7.
$$\begin{array}{r} 356 \\ +16 \\ \hline \end{array}$$

8.
$$\begin{array}{r} 268 \\ +\ 23 \\ \hline \end{array}$$

9.
$$\begin{array}{r} 739 \\ +25 \\ \hline \end{array}$$

10.
$$\begin{array}{r} 826 \\ +\ 47 \\ \hline \end{array}$$

Name: _____

1.
$$450$$
$$+286$$

2.
$$217$$
$$+746$$

3.
$$238$$
$$+635$$

4.
$$524$$
$$+449$$

5.
$$653$$
$$+276$$

6.
$$775$$
$$+215$$

7.
$$544$$
$$+138$$

8.
$$169$$
$$+826$$

9.
$$229$$
$$+267$$

10.
$$173$$
$$+645$$

1.
$$352$$
$$+479$$

2.
$$278$$
$$+645$$

3.
$$139$$
$$+576$$

4.
$$479$$
$$+342$$

5.
$$554$$
$$+298$$

6.
$$674$$
$$+189$$

7.
$$548$$
$$+369$$

8.
$$263$$
$$+677$$

9.
$$258$$
$$+267$$

10.
$$458$$
$$+356$$

Name: _____

1.
$$7,452 + 458$$

2.
$$4,249 + 658$$

3.
$$5,138 + 356$$

4.
$$1,526 + 267$$

5.
$$3,654 + 677$$

6.
$$3,977 + 263$$

7.
$$2,845 + 548$$

8.
$$8,664 + 369$$

9.
$$6,721 + 674$$

10.
$$7,253 + 189$$

Name: _____

1.
```
  5,351
+   298
```

2.
```
  9,118
+   754
```

3.
```
  3,137
+   342
```

4.
```
  6,425
+   479
```

5.
```
  7,553
+   576
```

6.
```
  4,876
+   139
```

7.
```
  1,744
+   645
```

8.
```
  3,563
+   278
```

9.
```
  2,620
+   479
```

10.
```
  8,252
+   352
```

Name: _____

1.
$$4,450$$
$$+5,253$$

2.
$$2,217$$
$$+3,729$$

3.
$$7,664$$
$$+1,654$$

4.
$$1,524$$
$$+2,845$$

5.
$$2,653$$
$$+6,452$$

6.
$$5,775$$
$$+3,138$$

7.
$$1,844$$
$$+4,249$$

8.
$$2,163$$
$$+1,526$$

9.
$$6,729$$
$$+2,260$$

10.
$$4,353$$
$$+1,645$$

1.
$$30,450 \\ + \underline{7,253}$$

2.
$$55,217 \\ + \underline{4,499}$$

3.
$$14,399 \\ + \underline{6,721}$$

4.
$$81,524 \\ + \underline{9,654}$$

5.
$$63,653 \\ + \underline{8,664}$$

6.
$$20,775 \\ + \underline{7,452}$$

7.
$$51,844 \\ + \underline{3,653}$$

8.
$$42,163 \\ + \underline{9,844}$$

9.
$$76,729 \\ + \underline{8,363}$$

10.
$$24,353 \\ + \underline{7,249}$$

Name: _____

1.
$$\begin{array}{r} 76{,}729 \\ +12{,}548 \\ \hline \end{array}$$

2.
$$\begin{array}{r} 48{,}309 \\ +37{,}751 \\ \hline \end{array}$$

3.
$$\begin{array}{r} 57{,}129 \\ +23{,}643 \\ \hline \end{array}$$

4.
$$\begin{array}{r} 19{,}318 \\ +71{,}653 \\ \hline \end{array}$$

5.
$$\begin{array}{r} 72{,}624 \\ +19{,}345 \\ \hline \end{array}$$

6.
$$\begin{array}{r} 59{,}564 \\ +30{,}857 \\ \hline \end{array}$$

7.
$$\begin{array}{r} 33{,}545 \\ +26{,}562 \\ \hline \end{array}$$

8.
$$\begin{array}{r} 46{,}381 \\ +13{,}736 \\ \hline \end{array}$$

9.
$$\begin{array}{r} 19{,}518 \\ +42{,}455 \\ \hline \end{array}$$

10.
$$\begin{array}{r} 33{,}453 \\ +16{,}657 \\ \hline \end{array}$$

1.
$$130,454$$
$$+ 69,546$$

2.
$$455,217$$
$$+ 25,941$$

3.
$$714,238$$
$$+ 73,834$$

4.
$$386,524$$
$$+ 16,447$$

5.
$$563,657$$
$$+ 38,334$$

6.
$$220,775$$
$$+ 69,443$$

7.
$$659,844$$
$$+ 45,174$$

8.
$$807,195$$
$$+ 57,826$$

9.
$$376,729$$
$$+ 28,260$$

10.
$$584,359$$
$$+ 32,645$$

1.   572,479
    +402,548

2.   158,268
    +751,751

3.   755,937
    +153,640

4.   317,718
    +247,653

5.   280,629
    +109,845

6.   269,567
    +125,234

7.   457,433
    +366,562

8.   346,256
    + 353,746

9.   159,518
    +378,349

10.   272,742
    +606,657

Name: _____

1.   5,145,459
      + 869,536

2.   7,455,517
      + 570,744

3.   3,784,238
      + 273,639

4.   1,381,584
      + 656,447

5.   4,563,653
      + 451,334

6.   8,279,775
      + 569,228

7.   2,654,844
      + 145,834

8.   6,842,164
      + 557,836

9.   4,276,729
      + 328,262

10.   1,524,359
       + 939,645

Unit 14 - Breaking Down a Problem

Breaking a problem down into smaller parts can help to make adding large numbers easier. We call this breaking down a problem or using partial sums to add.

You can add 6,723 + 2,262 using the math skills you have already learned or you can break the numbers down into smaller parts to make solving the problem easier.

$$6,723$$
$$+2,262$$

First, you will decompose the number into thousands, hundreds, tens, and ones.

	Thousands	Hundreds	Tens	Ones
	6,000	700	20	3
+	2,000	200	60	2

Now add each column starting with the ones and moving to the left toward the thousands.

	Thousands	Hundreds	Tens	Ones
	6,000	700	20	3
+	2,000	200	60	2
	8,000	900	80	5

Now, you can put the numbers back together to get the solution to the problem

$$6,723$$
$$+2,262$$
$$8,985$$

Name: _____

Decompose each number to help you solve the problem.

1.
22,849
+ 37,140

Ten Thousands	Thousands	Hundreds	Tens	Ones

2.
61,328
+ 7,551

Ten Thousands	Thousands	Hundreds	Tens	Ones

3.
46,120
+ 23,637

Ten Thousands	Thousands	Hundreds	Tens	Ones

4.
82,401
+ 3,228

Ten Thousands	Thousands	Hundreds	Tens	Ones

Name: _____

Decompose each number to help you solve the problem.

1.
    34,452
  + 2,311

Ten Thousands	Thousands	Hundreds	Tens	Ones

2.
    27,212
  + 11,435

Ten Thousands	Thousands	Hundreds	Tens	Ones

3.
    61,282
  + 3,617

Ten Thousands	Thousands	Hundreds	Tens	Ones

4.
    42,324
  +36,164

Ten Thousands	Thousands	Hundreds	Tens	Ones

Name: _____

Decompose each number to help you solve the problem.

1.
45,563
+ 3,373

Ten Thousands	Thousands	Hundreds	Tens	Ones

2.
31,323
+ 32,546

Ten Thousands	Thousands	Hundreds	Tens	Ones

3.
72,313
+ 4,720

Ten Thousands	Thousands	Hundreds	Tens	Ones

4.
53,435
+21,212

Ten Thousands	Thousands	Hundreds	Tens	Ones

Name: _____

Decompose each number to help you solve the problem.

1. | Ten Thousands | Thousands | Hundreds | Tens | Ones |
23,341
+ 1,250

2. | Ten Thousands | Thousands | Hundreds | Tens | Ones |
28,101
+ 51,546

3. | Ten Thousands | Thousands | Hundreds | Tens | Ones |
72,393
+ 4,605

4. | Ten Thousands | Thousands | Hundreds | Tens | Ones |
51,574
+26,214

Name: _____

Decompose each number to help you solve the problem.

1.

45,563
+ 3,326

Ten Thousands	Thousands	Hundreds	Tens	Ones

2.

38,232
+ 31,435

Ten Thousands	Thousands	Hundreds	Tens	Ones

3.

72,373
+24,606

Ten Thousands	Thousands	Hundreds	Tens	Ones

4.

53,435
+26,264

Ten Thousands	Thousands	Hundreds	Tens	Ones

Unit 15 - Rounding

Another way to make solving difficult problems easier is rounding. Rounding is a way of simplifying numbers to make them easier to understand.

You can use rounding whenever you do not need to know the exact answer. Solving a problem without the exact answer is called estimating. You can estimate whenever the problem you are solving can be answered with a close number instead of the exact number.

Example: If you are feeding 48 peanuts to the squirrel in your yard, you do not need to know the exact number of peanuts you give to him. You can say that you gave the squirrel 50 peanuts and still be accurate. This is called an estimate of the number of peanuts you fed to the squirrel.

How do you know that 48 peanuts rounds to 50?

You will use place value when you are rounding numbers.   First, you need to find the round off digit. We are rounding the number of peanuts to the nearest 10 so the number in the 10's place is the rounding number.

Tens   Ones

<u>4</u>   8

Next, we look at the number that is immediately to the right of the rounding number. That is the round off digit. In this example, the round off digit is the 8.

Tens   **Ones**

4   **<u>8</u>**

Next, you need to decide what to do with the 8. We round numbers to the nearest 10's place. 5 is the middle of the number line. If a number is 5 or more it is closer to the 10's so we round up. If a number is less than 5 it is closer to the 0 so we round down.

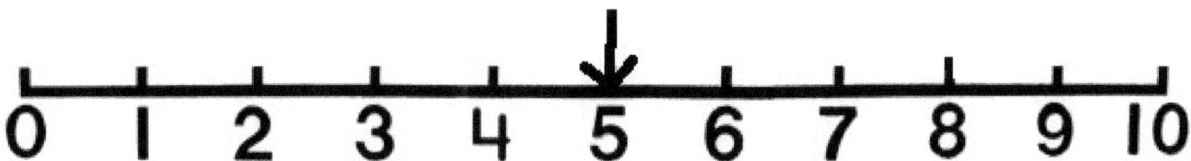

We need to round 48 to the nearest 10 place. 8 is our rounding off number.

Tens **Ones**

4   **8**

```
  0   1   2   3   4   5   6   7   8   9   10
```

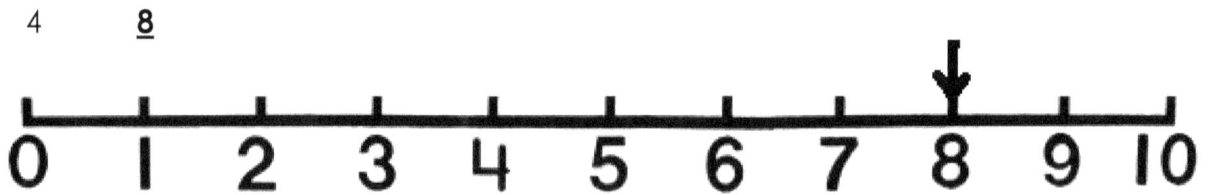

Now you will decide if the 8 is closer to the 10 or the 0.

8 is closer to the 10 so we will round the number 8 up to the next 10.

Our number was 48 so rounding up to the next 10 makes our rounded number 50.

Example: You fed 48 peanuts to the squirrels and your sister fed another 27 peanuts
            to the squirrels. About how many peanuts did you feed the squirrels in all?

```
    4    8    Rounds To          5    0
+   2    7    Rounds To      +   3    0
_____                  _____
                                 8    0
```

You fed the squirrels about 80 peanuts in all.

"About" is a clue word that tells you that you can answer the problem with an estimate instead of an exact calculation.

Name: _____

Use the number line to practice rounding these numbers to the nearest 10.

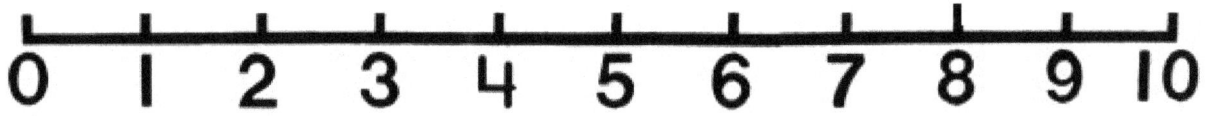

```
├──┼──┼──┼──┼──┼──┼──┼──┼──┼──┤
0  1  2  3  4  5  6  7  8  9  10
```

1.   Tens   Ones   2.   Tens   Ones
    29   51

    _____   _____

3.   Tens   Ones   4.   Tens   Ones
    43   18

    _____   _____

5.   Tens   Ones   6.   Tens   Ones
    24   35

    _____   _____ -
                                              _____

Name: _____

Use the number line to practice rounding these numbers to the nearest 10.

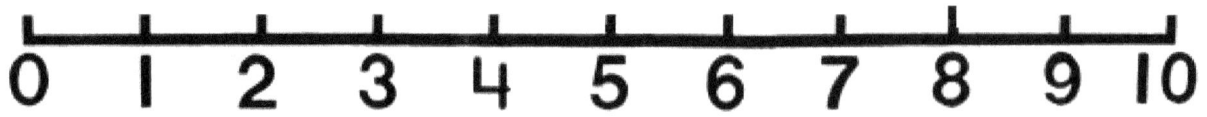

```
├──┼──┼──┼──┼──┼──┼──┼──┼──┼──┤
0   1   2   3   4   5   6   7   8   9  10
```

1.   Tens   Ones   2.   Tens   Ones

   57   32

   _____   _____

3.   Tens   Ones   4.   Tens   Ones

                                     16

   78

   _____   _____

5.   Tens   Ones   6.   Tens   Ones

   91   29

   _____   _____

Name: _____

Use the number line to practice rounding these numbers to the nearest 10.

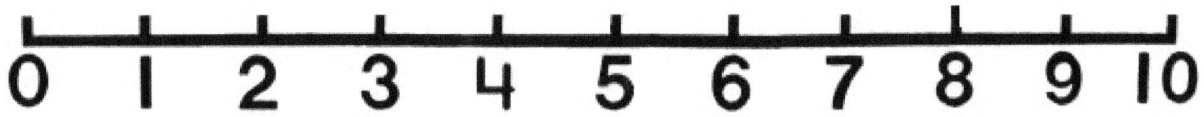

```
|——|——|——|——|——|——|——|——|——|——|
0   1   2   3   4   5   6   7   8   9  10
```

1.   Tens   Ones   2.   Tens   Ones
   48   13

   _____   _____

3.   Tens   Ones   4.   Tens   Ones
                                     25
   59

   _____   _____

5.   Tens   Ones   6.   Tens   Ones
   82   67

   _____   _____

Adding by Rounding to 10

A number line can also help you to add numbers when you do not need to know the exact answer.

"About" is a clue word that tells you that you can answer the problem with an estimate instead of an exact calculation.

Example: If you and your friend are helping to pick up sticks to build a fire, you do not need to know exactly how many sticks you collected. Your Dad told you that you need about 50 sticks to build a fire. If you collected 27 sticks and your friend collected 21 sticks, you can estimate to decide if you have enough to build the fire.

First, round each number to the nearest 10.

27 rounds up to 30 because the 7 is closer to 10 than to 0.

21 rounds down to 20 because the 1 is closer to 0 than to 10.

$$\begin{array}{r} 27 \\ + 21 \\ \hline \end{array} \quad \begin{array}{l} \text{Rounds To} \\ \text{Rounds To} \end{array} \quad \begin{array}{r} 30 \\ + 20 \\ \hline \end{array}$$

Now add the numbers

$$\begin{array}{r} 27 \\ + 21 \\ \hline \end{array} \quad \begin{array}{l} \text{Rounds To} \\ \text{Rounds To} \end{array} \quad \begin{array}{r} 30 \\ + 20 \\ \hline 50 \end{array}$$

30 + 20 = 50

Name: _____

Use the number line to help you round and solve each problem

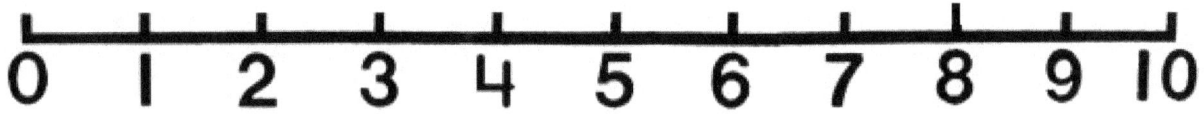

```
 |——|——|——|——|——|——|——|——|——|——|
 0  1  2  3  4  5  6  7  8  9  10
```

1.   46 Rounds To   2.   71 Rounds To
    + 32 Rounds To   + _____   + 18 Rounds To   + _____

3.   34 Rounds To   4.   16 Rounds To
    + 35 Rounds To   + _____   + 22 Rounds To   + _____

5.   57 Rounds To   6.   63 Rounds To
    + 32 Rounds To   + _____   + 34 Rounds To   + _____

7.   28 Rounds To   8.   75 Rounds To
    + 41 Rounds To   + _____   + 12 Rounds To   + _____

Name: _____

Use the number line to help you round and solve each problem

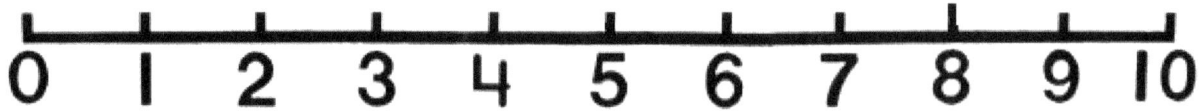

```
0   1   2   3   4   5   6   7   8   9   10
```

1.   67 Rounds To
   + 21 Rounds To + _____

2.   82 Rounds To
   + 14 Rounds To + _____

3.   45 Rounds To
   + 24 Rounds To + _____

4.   27 Rounds To
   + 11 Rounds To + _____

5.   27 Rounds To
   + 28 Rounds To + _____

6.   74 Rounds To
   + 23 Rounds To + _____

7.   39 Rounds To
   + 41 Rounds To + _____

8.   54 Rounds To
   + 22 Rounds To + _____

Name: _____

Use the number line to help you round and solve each problem

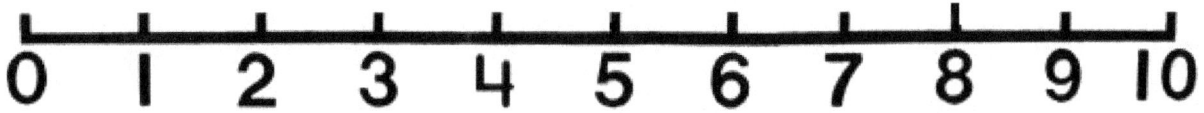

```
├──┼──┼──┼──┼──┼──┼──┼──┼──┼──┤
0   1   2   3   4   5   6   7   8   9  10
```

1.　　81　Rounds To
　　+ 12　Rounds To　+ _____

2.　　73　Rounds To
　　+ 24　Rounds To　+ _____

3.　　56　Rounds To
　　+ 33　Rounds To　+ _____

4.　　38　Rounds To
　　+ 22　Rounds To　+ _____

5.　　48　Rounds To
　　+ 37　Rounds To　+ _____

6.　　85　Rounds To
　　+ 14　Rounds To　+ _____

7.　　47　Rounds To
　　+ 28　Rounds To　+ _____

8.　　65　Rounds To
　　+ 15　Rounds To　+ _____

Rounding to 100

Rounding becomes more important when you are working with larger numbers. Adding 754 + 235 is a problem that you have the skills to solve. If you do not need to know the exact answer then you can make solving the problem easier by rounding.

You can round larger numbers using the same processes you learned when rounding tens and units.

First, decide which number is the rounding number.

You will always round the number immediately to the right of the rounding number.

Rounding 176 to the nearest 10 means that the 7 is the rounding number.

176		
hundreds	**tens**	ones
1	**7**	6

The number to the right of the 7 is a 6.

176		
hundreds	tens	**ones**
1	7	**6**

6 rounds up because it is closer to the 10 than to the 0 on the number line.

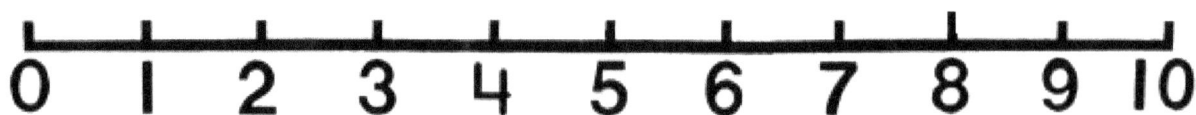

When we round up, we add the 10 that the 6 becomes to the rounding number making the answer 180.

Rounding 176 to the nearest 100 means that the 1 is the rounding number.

176		
hundreds	tens	ones
<u>1</u>	7	6

The number to the right of the 1 is a 7.

176		
hundreds	**tens**	ones
1	<u>7</u>	6

7 rounds up because it is closer to the 10 than to the 0 on the number line.

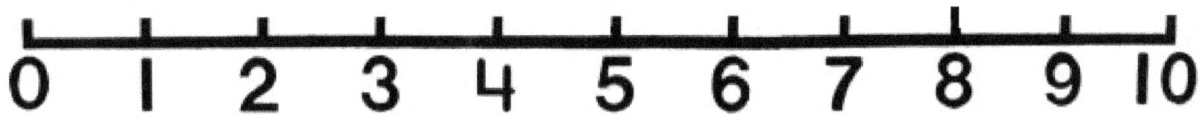

When we round up, we add the 100 that the 7 becomes to the rounding number making the answer 200.

Name: _____

Use the number line to practice rounding these numbers to the nearest 100.

```
|---|---|---|---|---|---|---|---|---|---|
0   1   2   3   4   5   6   7   8   9   10
```

1.　　　Hundreds　Tens　Ones　　2.　　　　Hundreds　Tens　Ones

　257　　　　　　　　　　　　　632

　_____　　　　_____

3.　　　Hundreds　Tens　Ones　　4.　　　　Hundreds　Tens　Ones

　　　　　　　　　　　　　　116

　578

　_____　　　　_____

5.　　　Hundreds　Tens　Ones　　6.　　　Hundreds　Tens　Ones

　391　　　　　　　　　　　　　829

　_____　　　　_____

Name: _____

Use the number line to practice rounding these numbers to the nearest 100.

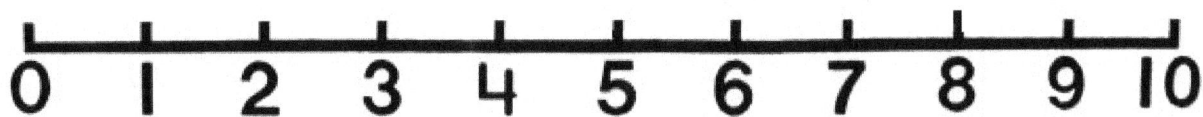

```
├──┼──┼──┼──┼──┼──┼──┼──┼──┼──┤
0   1   2   3   4   5   6   7   8   9  10
```

1.   Hundreds Tens Ones   2.   Hundreds Tens Ones
    368   275

    _____   _____

3.   Hundreds Tens Ones   4.   Hundreds Tens Ones
                                               584
    431
    _____   _____

5.   Hundreds Tens Ones   6.   Hundreds Tens Ones
    297   643

    _____   _____

Adding by Rounding to 100

A number line can also help you to add bigger numbers.

Example: Adding 675 + 237 is a problem that you have the skills to solve. If you do not need to now the exact answer and an estimate will do, then rounding can help to make solving the problem easier.

A lot of estimating problems will be story problems. When you see the word "about" it is a clue that you can use an estimate instead of calculating the exact answer.

First, round each number to the nearest 100.

6<u>7</u>5 rounds up to 700 because the 7 is closer to 10 than to 0.

2<u>3</u>7 rounds down to 200 because the 3 is closer to the 0 than to 10.

Remember that every number to the right of the one that you round turns into a 0.

$$
\begin{array}{r} 675 \\ + \ 237 \\ \hline \end{array}
\text{Rounds To}
\begin{array}{r} 700 \\ + \ 200 \\ \hline \end{array}
$$

Now add the numbers

$$
\begin{array}{r} 675 \\ + \ 237 \\ \hline \end{array}
\text{Rounds To}
\begin{array}{r} 700 \\ + \ 200 \\ \hline 900 \end{array}
$$

Adding 700 + 200 is easier than adding 675 + 237.

Name: _____

Use the number line to help you round and solve each problem

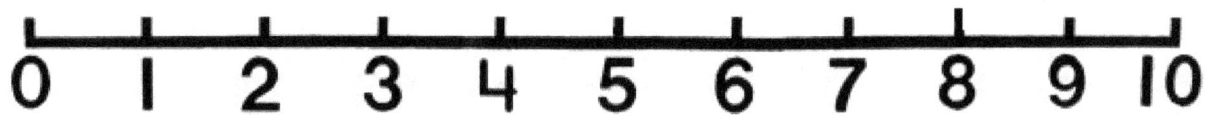

| | | | | | | | | | | |
0 1 2 3 4 5 6 7 8 9 10

1.   446   Rounds To   2.   671   Rounds To

   + 532   Rounds To   + _____   +118   Rounds To   + _____

3.   734   Rounds To   4.   316   Rounds To

   +235   Rounds To   + _____   +472   Rounds To   + _____

5.   157   Rounds To   6.   263   Rounds To

   + 132   Rounds To   + _____   +534   Rounds To   + _____

7.   528   Rounds To   8.   825   Rounds To

   + 341   Rounds To   + _____   +112   Rounds To   + _____

Name: _____

Use the number line to help you round and solve each problem

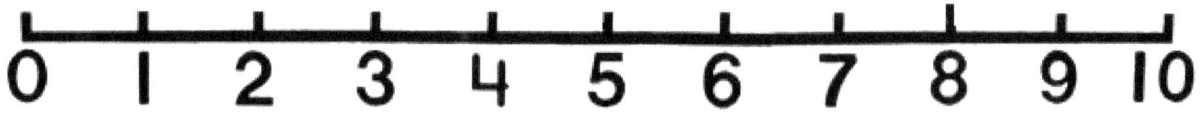

```
 |___|___|___|___|___|___|___|___|___|___|
 0   1   2   3   4   5   6   7   8   9   10
```

1. 746 Rounds To

    + 232 Rounds To +_____

2. 371 Rounds To

    +518 Rounds To +_____

3. 534 Rounds To

    +335 Rounds To +_____

4. 816 Rounds To

    +122 Rounds To +_____

5. 157 Rounds To

    +732 Rounds To +_____

6. 463 Rounds To

    +434 Rounds To +_____

7. 628 Rounds To

    +341 Rounds To +_____

8. 285 Rounds To

    +612 Rounds To +_____

Name: _____

Use the number line to help you round and solve each problem

```
 |  |  |  |  |  |  |  |  |  |  |
 0  1  2  3  4  5  6  7  8  9  10
```

1.   267   Rounds To   2.   182   Rounds To
    + 321   Rounds To   + _____   + 214   Rounds To   + _____

3.   445   Rounds To   4.   327   Rounds To
    + 524   Rounds To   + _____   + 311   Rounds To   + _____

5.   627   Rounds To   6.   574   Rounds To
    + 328   Rounds To   + _____   + 423   Rounds To   + _____

7.   839   Rounds To   8.   254   Rounds To
    + 141   Rounds To   + _____   + 122   Rounds To   + _____

Name: _____

Use the number line to help you round and solve each problem

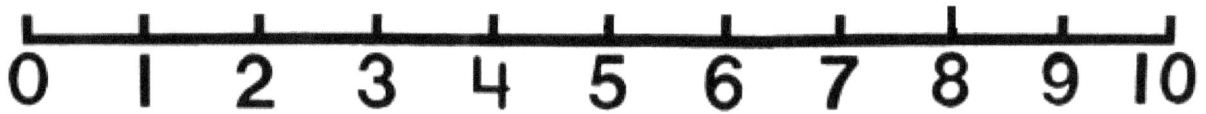

```
0  I  2  3  4  5  6  7  8  9  10
```

1. 581 Rounds To
   + 312 Rounds To + _____

2. 873 Rounds To
   + 124 Rounds To + _____

3. 556 Rounds To
   + 433 Rounds To + _____

4. 238 Rounds To
   + 722 Rounds To + _____

5. 748 Rounds To
   + 237 Rounds To + _____

6. 485 Rounds To
   + 514 Rounds To + _____

7. 147 Rounds To
   + 858 Rounds To + _____

8. 665 Rounds To
   + 325 Rounds To + _____

Rounding Bigger Numbers

Rounding becomes more important when you are working with larger numbers. Adding 74,254 + 23,735 is a problem that you have the skills to solve, but if you do not need to know the exact answer then you can make solving the problem easier by rounding.

You can round larger numbers using the same processes you learned when rounding tens and units.

First, decide which number is the rounding number.

You will always round the number immediately to the right of the rounding number.

Rounding 74,254 to the nearest 100 means that the **2** is the rounding number.

74,254					
Ten Thousands	Thousands	,	**hundreds**	tens	ones
7	4	,	**2**	5	4

The number to the right of the 2 is a 5.

74,254					
Ten Thousands	Thousands	,	hundreds	**tens**	ones
7	4	,	2	**5**	4

5 rounds up because it is closer to the 10 than to the 0 on the number line.

When we round up, the 5 becomes a 0 and the 1 form the tens is added to the 100's column. Remember, every number to the right of the rounding number becomes a 0

74,254 rounded to the nearest 100 = 74,300

Rounding 74,254 to the nearest 1,000 means that the 4 is the rounding number.

74,254					
Ten Thousands	**Thousands**	,	hundreds	tens	ones
7	**4**	,	2	5	4

The number to the right of the 4 is a 2.

74,254					
Ten Thousands	Thousands	,	**hundreds**	tens	ones
7	4	,	**2**	5	4

2 rounds down because it is closer to the 0 than to the 10 on the number line.

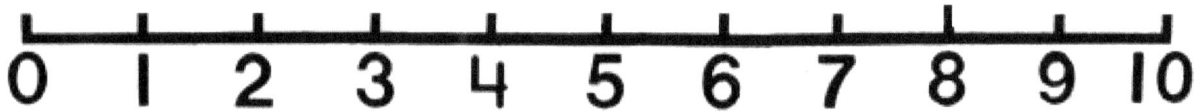

When we round down, we do not add anything to the thousands place because our number rounded to 0. Remember, every number to the right of the rounding number becomes a 0.

74,254 rounded to the nearest 1,000 = 74,000

Rounding 23,735 to the nearest 100 means that the 7 is the rounding number.

23,735					
Ten Thousands	Thousands	,	**hundreds**	tens	ones
2	3	,	**7**	3	5

The number to the right of the 7 is a 3.

23,735

Ten Thousands	Thousands	,	hundreds	**tens**	ones
2	3	,	7	<u>3</u>	5

3 rounds down because it is closer to the 0 than to the 10 on the number line.

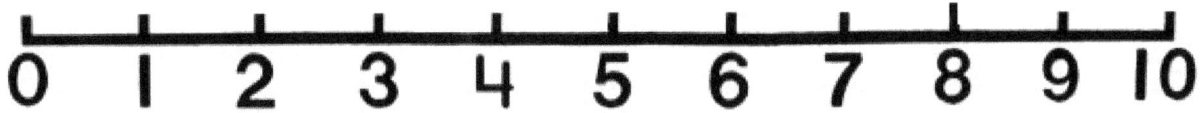

When we round down, the 3 becomes a 0 and we do not add anything to the 100 place. Remember, every number to the right of the rounding number becomes a 0

23,735 rounded to the nearest 100 = 23,700

Rounding 23,735 to the nearest 1,000 means that the 3 is the rounding number.

23,735					
Ten Thousands	**Thousands**	,	hundreds	tens	ones
2	<u>3</u>	,	7	3	5

The number to the right of the 3 is a 7.

23,735					
Ten Thousands	Thousands	,	**hundreds**	tens	ones
2	3	,	<u>7</u>	3	5

7 rounds up because it is closer to the 10 than to the 0 on the number line.

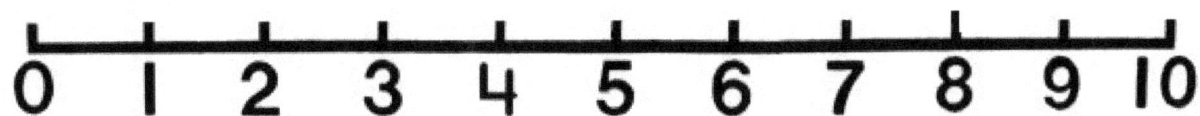

When we round up, we add the 1,000 that the 7 became to the thousands place. Remember, every number to the right of the rounding number becomes a 0.

23,735 rounded to the nearest 1,000 = 24,000

   74,254   Rounds To   74,000

 +23,735   Rounds To   +24,000
                        98,000

Name: _____

Use the number line to practice rounding these numbers to the nearest 100.

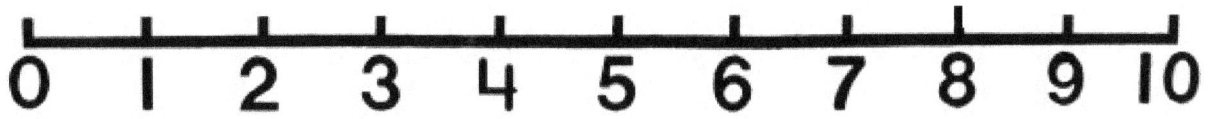

```
├──┼──┼──┼──┼──┼──┼──┼──┼──┼──┤
0   1   2   3   4   5   6   7   8   9  10
```

1.

48,257

_____

2.

71,864

_____

3.

93,641

_____

4.

23,864

_____

Name: _____

Use the number line to practice rounding these numbers to the nearest 100

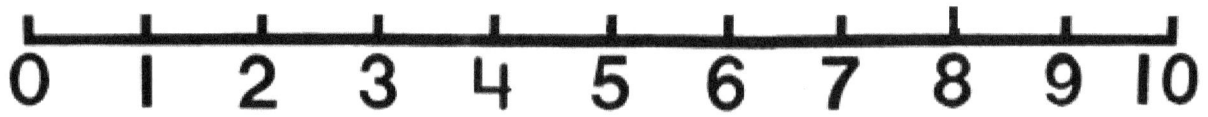

```
├──┼──┼──┼──┼──┼──┼──┼──┼──┼──┤
0  1  2  3  4  5  6  7  8  9  10
```

1.

87,727

_____

2.

74,318

_____

3.

12,972

_____

4.

35,818

_____

Name: _____

Use the number line to practice rounding these numbers to the nearest 1,000.

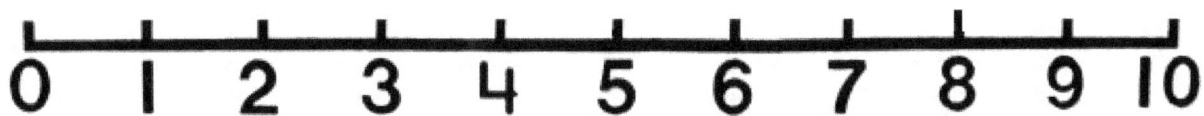

```
├──┼──┼──┼──┼──┼──┼──┼──┼──┼──┤
0   1   2   3   4   5   6   7   8   9  10
```

1.

91,244

_____

2.

18,635

_____

3.

24,581

_____

4.

46,427

_____

Name: _____

Use the number line to practice rounding these numbers to the nearest 1,000.

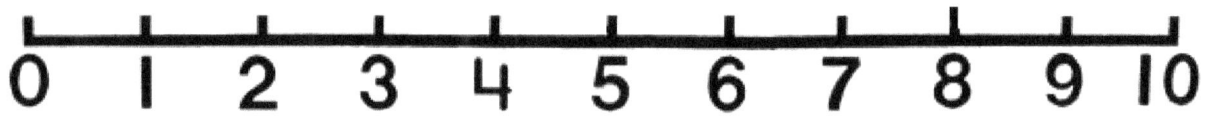

```
 |——|——|——|——|——|——|——|——|——|——|
 0  1  2  3  4  5  6  7  8  9  10
```

1.

   35,818

_____

2.

   87,727

_____

3.

   74,318

_____

4.

   94,611

_____

Name: _____

Use the number line to practice rounding these numbers to the nearest 10,000.

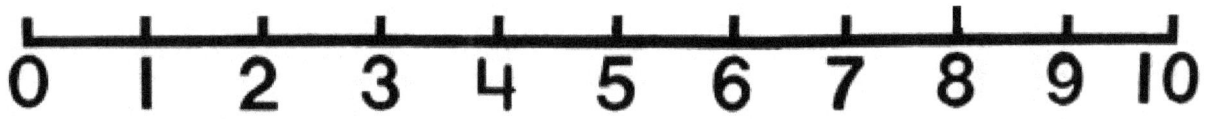

```
|——|——|——|——|——|——|——|——|——|——|
0   1   2   3   4   5   6   7   8   9  10
```

1.

48,257

_____

2.

71,864

_____

3.

93,641

_____

4.

17,916

_____

Name: _____

Use the number line to practice rounding these numbers to the nearest 10,000

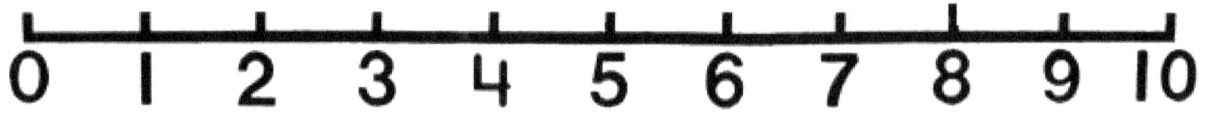

```
|---|---|---|---|---|---|---|---|---|---|
0   1   2   3   4   5   6   7   8   9  10
```

1.

35,818

_____

2.

87,727

_____

3.

74,318

_____

4.

23,144

_____

Name: _____

Use the number line to practice rounding these numbers to the nearest 100,000.

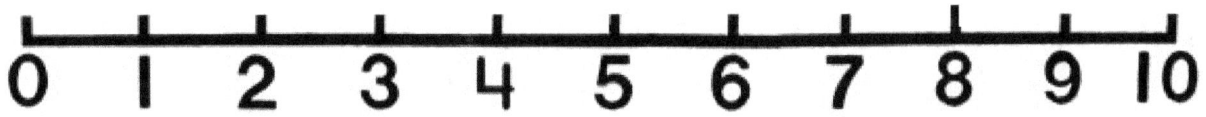

```
├──┼──┼──┼──┼──┼──┼──┼──┼──┼──┤
0   1   2   3   4   5   6   7   8   9  10
```

1.

   391,244

   _____

2.

   618,635

   _____

3.

   824,581

   _____

4.

   145,967

   _____

Name: _____

Use the number line to practice rounding these numbers to the nearest 100,000.

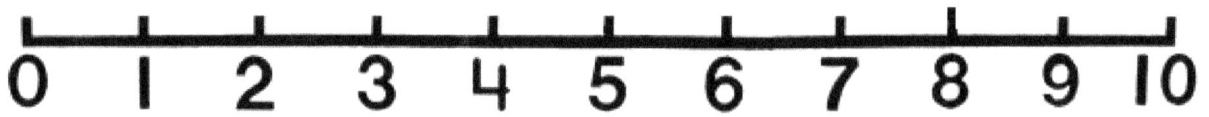

```
|---|---|---|---|---|---|---|---|---|---|
0   1   2   3   4   5   6   7   8   9   10
```

1.

   435,818

   _____

2.

   287,727

   _____

3.

   574,318

   _____

4.

   792,999

   _____

Name: _____

Use the number line to practice rounding these numbers to the nearest 1,000,000.

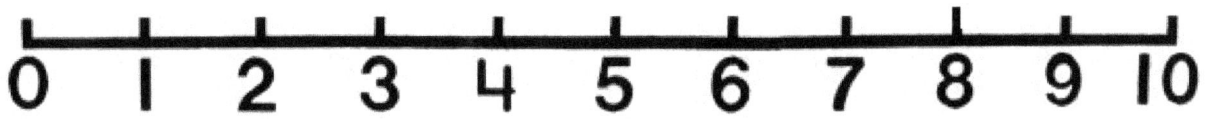

```
 |——|——|——|——|——|——|——|——|——|——|
 0   1   2   3   4   5   6   7   8   9  10
```

1.

   7,391,244

_____

2.

   4,618,635

_____

3.

   5,824,581

_____

4.

   2,986,555

_____

Name: _____

Use the number line to practice rounding these numbers to the nearest 1,000,000.

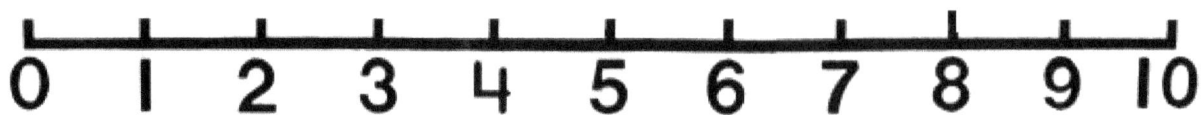

```
 ├──┼──┼──┼──┼──┼──┼──┼──┼──┼──┤
 0  1  2  3  4  5  6  7  8  9  10
```

1.

    9,435,818

_____

2.

    1,287,727

_____

3.

    3,574,318

_____

4.

    6,784,232

_____

Adding Bigger Numbers by Rounding

A number line can also help you to add bigger numbers.

Example: Adding 1,675,231 + 3,213,464 is a problem that you have the skills to solve, but if you do not need to now the exact answer and an estimate will do, then rounding can help to make solving the problem easier.

First, round each number to the nearest 100,000.

1,675,231rounds up to 1,700,000

3,213,464 rounds down to 3,200,000

Remember that every number to the right of the one that you round turns into a 0.

$$
\begin{array}{ll}
1,675,231 & \text{Rounds To} \\
\underline{+\ 3,213,464} & \text{Rounds To}
\end{array}
\qquad
\begin{array}{l}
1,700,000 \\
\underline{+\ 3,200,000}
\end{array}
$$

Now add the numbers

$$
\begin{array}{ll}
1,675,231 & \text{Rounds To} \\
\underline{+\ 3,213,464} & \text{Rounds To}
\end{array}
\qquad
\begin{array}{l}
1,700,000 \\
\underline{+\ 3,200,000} \\
4,900,000
\end{array}
$$

Adding 1,700,000 + 3,200,000 is easier than adding 1,675,231 + 3,213,464.

Name: _____

Use the number line to help you round each number to the nearest 10,000 and solve each problem.

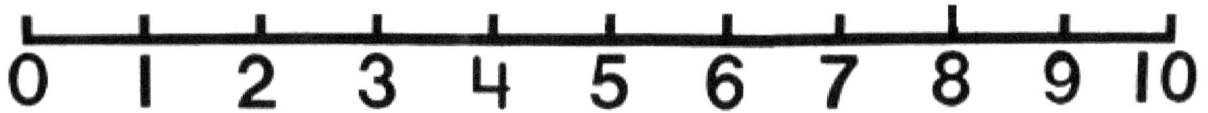

```
├──┼──┼──┼──┼──┼──┼──┼──┼──┼──┤
0   1   2   3   4   5   6   7   8   9  10
```

1. 24,446   Rounds To

    + 13,532   Rounds To   +_____

2. 36,734   Rounds To

    + 51,235   Rounds To   +_____

3. 73,157   Rounds To

    + 14,132   Rounds To   +_____

4. 45,528   Rounds To

    + 41,341   Rounds To   +_____

Name: _____

Use the number line to help you round each number to the nearest 10,000 and solve each problem.

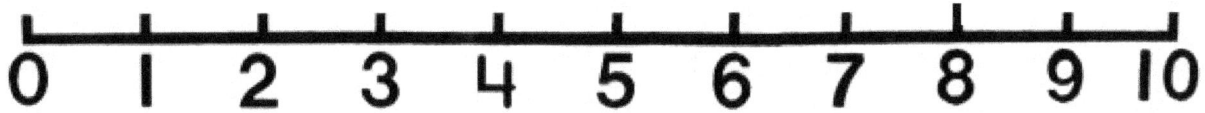

```
├──┼──┼──┼──┼──┼──┼──┼──┼──┼──┤
0  1  2  3  4  5  6  7  8  9  10
```

1.   51,999   Rounds To

   + 26,318   Rounds To   +_____

2.   16,823   Rounds To

   + 40,124   Rounds To   +_____

3.   72,046   Rounds To

   + 13,021   Rounds To   +_____

4.   35,417   Rounds To

   + 33,230   Rounds To   +_____

Name: _____

Use the number line to help you round each number to the nearest 100,000 and solve each problem.

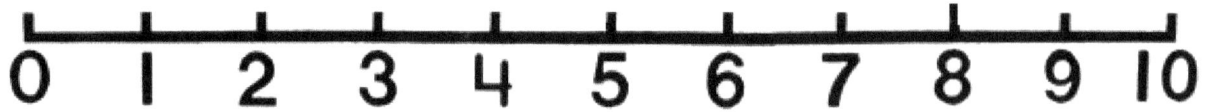

```
├──┼──┼──┼──┼──┼──┼──┼──┼──┼──┤
0   1   2   3   4   5   6   7   8   9  10
```

1. 637,523   Rounds To

    + 122,518   Rounds To   +_____

2. 146,318   Rounds To

    + 821,261   Rounds To   +_____

3. 318,921   Rounds To

    + 534,032   Rounds To   +_____

4. 725,648   Rounds To

    + 241,350   Rounds To   +_____

Use the number line to help you round each number to the nearest 100,000 and solve each problem.

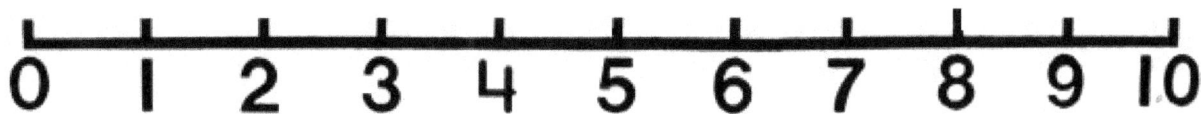

```
├──┼──┼──┼──┼──┼──┼──┼──┼──┼──┤
0   1   2   3   4   5   6   7   8   9   10
```

1.　251,876　　Rounds To

　　+ 634,123　　Rounds To　　+ _____

2.　456,871　　Rounds To

　　+ 433,026　　Rounds To　　+ _____

3.　172,369　　Rounds To

　　+ 123,520　　Rounds To　　+ _____

4.　415,724　　Rounds To

　　+ 484,234　　Rounds To　　+ _____

Name: _____

Use the number line to help you round each number to the nearest 1,000,000 and solve each problem.

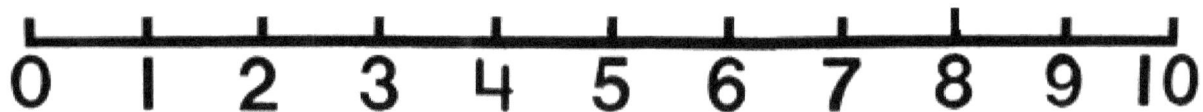

```
├──┼──┼──┼──┼──┼──┼──┼──┼──┼──┤
0   1   2   3   4   5   6   7   8   9  10
```

1. 3,526,412   Rounds To

    +3,011,407   Rounds To   + _____

2. 2,035,207   Rounds To

    +2,740,150   Rounds To   + _____

3. 4,207,810   Rounds To

    + 4,423,021   Rounds To   + _____

4. 7,614,537   Rounds To

    + 1,130,240   Rounds To   + _____

Name: _____

Use the number line to help you round each number to the nearest 1,000,000 and solve each problem.

```
 |  |  |  |  |  |  |  |  |  |  |
 0  1  2  3  4  5  6  7  8  9  10
```

1.   2,341,765   Rounds To

    +7,434,123   Rounds To   + _____

2.   6,445,760   Rounds To

    +3,322,015   Rounds To   + _____

3.   1,761,258   Rounds To

    + 1,112,411   Rounds To   + _____

4.   4,414,613   Rounds To

    + 4,584,285   Rounds To   + _____

Unit 16 - Adding Three Numbers – Three Addends

Sometimes you will have a problem to solve that has three or more numbers to add.

Example: 613 + 285 + 101.

You will line up these numbers just like you do with any addition problem.

```
 613
 285
+102
```

Add the ones and regroup if you can.

```
  1
 613
 285
+102
   0
```

Finish the math that you need to solve the problem.

```
 1 1
 613
 285
+102
1,000
```

You can add numbers in any order.

613 + 285 + 101 = 285 + 101 + 613 = 101 + 613 + 285

No matter what order you put the numbers in, the answer is the same.

Name: _____

Line up the numbers and then add solve each problem.

1.
$17 + 14 + 5 =$ _____

_____

2.
$21 + 15 + 4 =$ _____

_____

3.
$15 + 11 + 8 =$ _____

_____

4.
$46 + 12 + 7 =$ _____

_____

5.
$72 + 13 + 5 =$ _____

_____

Name: _____

Line up each problem and then solve it.

1.

68 + 12 + 91 = _____

_____

2.

328 + 116 + 422 = _____

_____

3.

612 + 315 + 187 = _____

_____

4.

417 + 169 + 312 = _____

_____

5.

180 + 273 + 511 = _____

_____

Name: _____

Line up each problem and then solve it.

1.

187 + 624 + 235 = _____

_____

2.

318 + 275 + 240 = _____

_____

3.

551 + 219 + 438 = _____

_____

4.

532 + 261 + 57 = _____

_____

5.

617 + 83 + 215 = _____

_____

Name: _____

Line up each problem and then solve it.

1.

$1,068 + 212 + 391 =$ _____

_____

2.

$2,328 + 1,116 + 4,422 =$ _____

_____

3.

$7,612 + 315 + 1,187 =$ _____

_____

4.

$4,417 + 2,169 + 1,312 =$ _____

_____

5.

$3,180 + 5,273 + 511 =$ _____

_____

Name: _____

Line up each problem and then solve it.

1.

$5,187 + 1,624 + 235 =$ _____

_____

2.

$3,318 + 2,275 + 1,240 =$ _____

_____

3.

$6,551 + 219 + 2,438 =$ _____

_____

4.

$2,532 + 4,261 + 1,957 =$ _____

_____

5.

$8,617 + 983 + 215 =$ _____

_____

Unit 17 - Real World Story Problems

Sometimes you will solve math problems that look different from what you are used to. These math problems will use words to describe a real life situation. These are called story problems or real world problems. In life, you will use your math skills to solve many real world problems so it is important that you learn the steps you must use to solve these types of problems.

You can use some simple steps to help make solving real world story problems easier.

- **Read the problem.**

   You must read the problem carefully to make certain you understand what the problem is asking.

- **Ignore extra information.**

   Some story problems will contain information that is not important to solving the problem. You should read the problem to decide what is important and what is not important to helping you to solve the problem. Cross out any information that is not important so that it does not confuse you as you solve the problem.

- **Think about the problem.**

   You should think about what the problem is asking you to solve.

- **Make a model of the problem.**

   Sometimes it is easiest to solve a story problem when you change the written information into pictures. Read the problem and decide if you can draw a picture to help you solve the problem.

- **Write the number sentence.**

   You should use all of the information you decided was important to write a number sentence. The number sentence will be the actual problem.

- **Solve the problem.**

- **Check your work.**

   You should compare your answer to what you think the problem is asking. Think about whether your answer makes sense when compared to the problem.

There are many steps to solving a story problem so you should always check your work to make sure you did not make a simple calculation mistake.

Example: Sydney has 2 cats and 1 dog. Her brother has 2 dogs and 1 cat. How many cats do the kids have in all?

Step 1:   Read the problem.

When you read this problem, you see the question "How many cats do the kids have in all?" This is the problem you are being asked to solve.

Step 2:   Ignore any extra information.

The question is about cats so the parts of the question that show how many dogs the kids have is just extra information that does not help you to solve the problem.

When we remove the extra information, we are left with the information that helps us solve the problem.

Sydney has 2 cats and her brother has 1 cat.

Step 3:   Make a model of the problem.

Sydney has

Stephen has

Step 4:   Think about the problem.

You need to decide what the problem is asking you to solve. This one is easy. The problem asks, "How many cats do the kids have in all?"

Step 5:   Write the number sentence.

2 + 1 =

Step 6:   Solve the problem.

2 + 1 = 3

Step 7:   Check your work.

When you check your work, you should think about the steps you took to solve the problem.

Does it make sense that the kids have 3 cats?

The question asks in all. In all usually tells us that we are working on an addition problem. Did you use addition to solve the problem?

Did you add correctly?

That problem was pretty easy to solve. You will use the same steps to solve harder story problems.

Example: Stephen was selling flowers to raise money for his class trip. He sold 27 flowers to his neighbors, 8 flowers to his Dad's co-workers and 19 flowers to his family. How many flowers did Stephen sell in all?

Step 1:   Read the problem.

When you read this problem, you see the question "How many flowers did Stephen sell in all?" This is the problem you are being asked to solve.

Step 2:   Ignore any extra information.

There is no extra information in this problem. Everything is important.

Step 3:   Make a model of the problem.

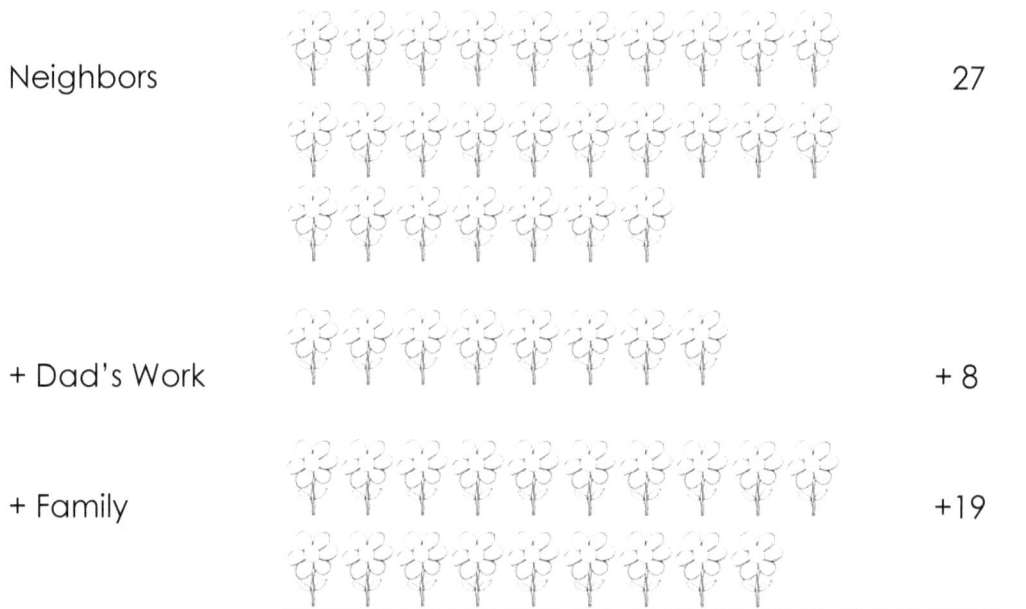

Neighbors		27
+ Dad's Work		+ 8
+ Family		+19

Step 4:   Think about the problem.

190

You need to decide what the problem is asking you to solve. The problem asks, "How many flowers did Stephen sell in all?"

Step 5:   Write the number sentence.

27 + 8 + 19

Step 6:   Solve the problem.

27 + 8 + 19 = 54

Step 7:   Check your work.

When you check your work, you should think about the steps you took to solve the problem.

Does it make sense that the Stephen sold 54 flowers?

The question asks in all. In all usually tells us that we are working on an addition problem. Did you use addition to solve the problem?

Did you add correctly?

Name: _____

Follow the steps to solve each story problem.

1. Stephen has 9 toy cars. His friend came over with 7 more toy cars and his racetrack. How many toy cars do the boys have in all?

Read the problem.
Ignore extra information.
Make a model of the problem.

Think about the problem.
Write the number sentence.

Solve the problem.
Check your work.

_____

2. Mom baked 24 chocolate chop cookies, 36 sugar cookies, and 12 cupcakes. How many cookies did she bake in all?

Read the problem.
Ignore extra information.
Make a model of the problem.

Think about the problem.
Write the number sentence.

Solve the problem.
Check your work.

_____

3. There are 9 flowers growing in the garden, I planted 4 more flowers.
   How many flowers are in the garden?

Read the problem.
Ignore extra information.
Make a model of the problem.

Think about the problem.
Write the number sentence.

Solve the problem.
Check your work.

---

4. Sarah's box has 15 crayons. Bill's box has 12 crayons. How many
   crayons are there in all?

Read the problem.
Ignore extra information.
Make a model of the problem.

Think about the problem.
Write the number sentence.

Solve the problem.
Check your work.

Name: _____

Follow the steps to solve each story problem.

1. Karen scored 51 points in the game and Steve scored 37 points. How many points did they score in all?

Read the problem.
Ignore extra information.
Make a model of the problem.

Think about the problem.
Write the number sentence.

Solve the problem.
Check your work.

_____

2. Susan read 27 pages of her book on Monday, 38 on Tuesday, and 2 pages on Wednesday. How many pages did she read in all?

Read the problem.
Ignore extra information.
Make a model of the problem.

Think about the problem.
Write the number sentence.

Solve the problem.
Check your work.

_____

3. Leslie has 118 trading cards and her sister has 96. How many trading cards do the girls have in all?

4. Penny has 41 girls and 27 boys in her karate class. Judy has 36 boys and 19 girls in her karate class. How many girls are taking in the classes in all?

5. Ben and Rick collected 112 tin cans on Friday, 97 tin cans on Saturday, and 131 tin cans on Sunday. How many tin cans did they collect in all?

Name: _____

Follow the steps to solve each story problem.

1. The library had a book sale. They sold 48 children's books, 126 non-fiction books, and 235 fiction books. How many books did they sell in all?

_____

2. Karen scored 51 points in the game and Steve scored 37 points. How many points did they score in all?

_____

3. The girls caught 11 trout, 16 bluegill and 7 catfish while we were on vacation. How many fish did the girls catch in all?

_____

Name: _____

Follow the steps to solve each story problem.

1. Mark delivered 41 newspapers, Sarah delivered 22 magazines, and Jose delivered 36 newspapers. How many newspapers did the children deliver in all?

2. 117 people rode the roller coaster, 96 people ate at the concession stand, and 137 people rode the bumper cars. How many people rode rides?

3. Our club got 25 new machines and 37 new members last month. We got 18 new machines and 31 new members this month. How many new members did our club get in all?

4. 17children were playing on the swings. There were 11 children on the slide. 12 more children went to ride the swings. How many children were riding the swings in all?

5. Our team won our game 27-14. It was 92 degrees outside so the coach took us for ice cream. 7 of us had chocolate cones, 9 had vanilla cones, and 5 ate yogurt cups. How many ice cream cones did we eat in all?

6. Our music department has 12 tubas, 19 violins, and 11 flutes. We have 21 music students. How many instruments does the music department have in all?

ANSWERS

Page 10	Page 12	Page 14	Page 16	Page 18
1. 4	1. 5	1. 6	1. 7	1. 8
2. 6	2. 7	2. 8	2. 9	2. 10
3. 3	3. 4	3. 5	3. 6	3. 7
4. 8	4. 9	4. 10	4. 10	4. 11
5. 7	5. 6	5. 7	5. 8	5. 9

Page 20	Page 22	Page 24	Page 26	Page 28
1. 9	1. 10	1. 11	1. 12	1. 8
2. 10	2. 11	2. 12	2. 13	2. 21
3. 8	3. 9	3. 10	3. 11	3. 15
4. 11	4. 12	4. 13	4. 14	4. 12
5. 12	5. 11	5. 14	5. 15	5. 23

Page 30	Page 31	Page 32	Page 33	Page 34
1. 15	1. 14	1. 6	1. 8	1. 12
2. 16	2. 11	2. 10	2. 10	2. 18
3. 12	3. 7	3. 13	3. 14	3. 9
4. 9	4. 8	4. 10	4. 13	4. 14

Page 35	Page 36	Page 37	Page 38
1. 11	1. 10	1. 15	1. 17
2. 8	2. 10	2. 14	2. 11
3. 12	3. 12	3. 10	3. 13
4. 7	4. 9	4. 14	4. 10

Page 51	Page 52	Page 53	Page 54	Page 55
1. 9	1. 10	1. 11	1. 12	1. 13
2. 5	2. 6	2. 7	2. 8	2. 9
3. 7	3. 8	3. 9	3. 10	3. 11
4. 2	4. 3	4. 4	4. 5	4. 6
5. 4	5. 5	5. 6	5. 7	5. 8
6. 3	6. 4	6. 5	6. 6	6. 7
7. 6	7. 7	7. 8	7. 9	7. 10
8. 8	8. 9	8. 10	8. 11	8. 12

Page 56	Page 57	Page 58	Page 59	Page 60
1. 14	1. 15	1. 16	1. 17	1. 18
2. 10	2. 11	2. 12	2. 13	2. 14
3. 12	3. 13	3. 14	3. 15	3. 16
4. 7	4. 8	4. 9	4. 10	4. 11
5. 9	5. 10	5. 11	5. 12	5. 13
6. 8	6. 9	6. 10	6. 11	6. 12
7. 11	7. 12	7. 13	7. 14	7. 15
8. 13	8. 14	8. 15	8. 16	8. 17

Page 61						Page 62					
1. 12	22	32	42	52	62	1. 12	22	32	42	52	62
2. 7	17	27	37	47	57	2. 14	24	34	44	54	64
3. 11	21	31	41	51	61	3. 14	24	34	44	54	64
						4. 10	20	30	40	50	60
						5. 12	22	32	42	52	62
						6. 15	25	35	45	55	65

Page 63						Page 64	Page 66	Page 67
1. 11	21	31	41	51	61	1. 23	1. 19	1. 22
2. 17	27	37	47	57	67	2. 19	2. 24	2. 21
3. 12	22	32	42	52	62	3. 20	3. 17	3. 18
4. 12	22	32	42	52	62	4. 23	4. 22	4. 24
5. 16	26	36	46	56	66	5. 21		
6. 18	28	38	48	58	68			

Page 68	Page 69	Page 72	Page 73	Page 74
1. 22	1. 22	1. 76	1. 66	1. 80
2. 18	2. 24	2. 57	2. 48	2. 62
3. 24	3. 19	3. 23	3. 34	3. 98
4. 20	4. 23	4. 25	4. 36	4. 83
		5. 37	5. 26	5. 37
		6. 42	6. 60	6. 30
		7. 38	7. 69	7. 58
		8. 69	8. 82	8. 92
		9. 34	9. 98	9. 98
		10. 50	10. 95	10. 71

Page 75	Page 76	Page 77	Page 78	Page 79
1. 95	1. 92	1. 67	1. 67	1. 66
2. 87	2. 74	2. 56	2. 56	2. 79
3. 45	3. 65	3. 48	3. 48	3. 86
4. 56	4. 60	4. 58	4. 58	4. 37
5. 57	5. 83	5. 59	5. 59	5. 87
6. 64	6. 97	6. 98	6. 98	6. 76
7. 54	7. 93	7. 87	7. 87	7. 65
8. 79	8. 82	8. 98	8. 98	8. 76
9. 64	9. 50	9. 98	9. 98	9. 96
10. 80	10. 89		10. 87	10. 99
			11. 88	11. 99
			12. 79	12. 98

Page 80	Page 81	Page 82	Page 83	Page 84
1. 66	1. 78	1. 456	1. 657	1. 566
2. 89	2. 89	2. 447	2. 586	2. 879
3. 88	3. 99	3. 498	3. 498	3. 896
4. 59	4. 79	4. 958	4. 958	4. 937
5. 87	5. 87	5. 659	5. 959	5. 897
6. 98	6. 98	6. 998	6. 998	6. 876
7. 87	7. 87	7. 587	7. 787	7. 965
8. 97	8. 98	8. 998	8. 998	8. 796
9. 96	9. 49	9. 998	9. 798	9. 996
10. 89	10. 89		10. 897	10. 999
11. 99	11. 79		11. 798	11. 999
12. 98	12. 98		12. 998	12. 898

Page 85	Page 86	Page 89	Page 90	Page 91
1. 696	1. 798	1. 17	1. 39	1. 659
2. 899	2. 989	2. 15	2. 38	2. 299
3. 788	3. 899	3. 19	3. 27	3. 499
4. 599	4. 799	4. 19	4. 29	4. 789
5. 897	5. 887	5. 18	5. 29	5. 498
6. 998	6. 798			
7. 887	7. 997	Page 92	Page 93	
8. 997	8. 898	1. 759	1. 549	
9. 996	9. 849	2. 998	2. 786	
10. 989	10. 889	3. 987	3. 737	
11. 999	11. 799	4. 759	4. 537	
12. 998	12. 988	5. 959	5. 737	

Page 95	Page 96	Page 97	Page 98	Page 99
1. 477	1. 387	1. 986	1. 788	1. 7,477
2. 249	2. 159	2. 958	2. 859	2. 4,249
3. 159	3. 169	3. 868	3. 669	3. 5,159
4. 558	4. 468	4. 967	4. 965	4. 1,558
5. 677	5. 587	5. 987	5. 989	5. 9,677
6. 989	6. 899	6. 998	6. 798	6. 3,989
7. 868	7. 778	7. 978	7. 896	7. 2,868
8. 699	8. 589	8. 989	8. 888	8. 8,699
9. 747	9. 657	9. 989	9. 889	9. 6,747
10. 287	10. 297	10. 998	10. 998	10. 7,287

Page 100	Page 101	Page 102	Page 103	Page 104
1. 5,387	1. 4,986	1. 2,779	1. 4,986	1. 9,687
2. 9,159	2. 7,958	2. 8,859	2. 7,958	2. 9,998
3. 3,169	3. 8,868	3. 5,669	3. 8,868	3. 8,778
4. 6,468	4. 1,967	4. 7,966	4. 1,967	4. 9,878
5. 7,587	5. 3,987	5. 1,969	5. 3,987	5. 9,877
6. 4,899	6. 5,998	6. 9,798	6. 5,998	6. 8,887
7. 1,778	7. 9,978	7. 3,995	7. 9,978	7. 8,984
8. 3,589	8. 2,989	8. 6,989	8. 2,989	8. 9,497
9. 2,657	9. 6,989	9. 9,859	9. 6,989	9. 6,969
10. 8,297	10. 4,998	10. 4,999	10. 4,998	10. 9,899

Page 105	Page 106	Page 107	Page 108	Page 109
1. 39,986	1. 84,779	1. 199,986	1. 974,779	1. 5,999,986
2. 59,958	2. 79,859	2. 479,958	2. 999,859	2. 7,979,958
3. 17,868	3. 78,669	3. 787,868	3. 868,669	3. 3,987,868
4. 87,967	4. 88,966	4. 397,967	4. 958,966	4. 1,997,967
5. 64,987	5. 99,969	5. 594,987	5. 989,969	5. 4,994,987
6. 29,998	6. 99,798	6. 289,998	6. 789,798	6. 8,789,998
7. 56,978	7. 89,995	7. 696,978	7. 799,995	7. 2,796,978
8. 49,989	8. 89,989	8. 899,989	8. 999,989	8. 6,999,989
9. 77,989	9. 59,859	9. 298,989	9. 889,859	9. 4,597,989
10. 26,998	10. 88,999	10. 556,998	10. 878,999	10. 1,956,998

Page 121	Page 122	Page 123	Page 124	Page 125
1. 92	1. 33	1. 71	1. 517	1. 371
2. 81	2. 64	2. 70	2. 251	2. 870
3. 65	3. 81	3. 52	3. 209	3. 682
4. 60	4. 62	4. 85	4. 618	4. 501
5. 83	5. 93	5. 72	5. 682	5. 184
6. 73	6. 90	6. 53	6. 992	6. 275
7. 61	7. 82	7. 71	7. 928	7. 582
8. 82	8. 91	8. 94	8. 719	8. 792
9. 54	9. 64	9. 61	9. 790	9. 600
10. 95	10. 73	10. 43	10. 347	10. 364

Page 126	Page 127	Page 128	Page 129	Page 130
1. 263	1. 736	1. 831	1. 7,910	1. 5,649
2. 583	2. 963	2. 923	2. 4,907	2. 9,872
3. 172	3. 873	3. 715	3. 5,494	3. 3,479
4. 682	4. 973	4. 821	4. 1,793	4. 6,904
5. 491	5. 929	5. 852	5. 4,331	5. 8,129
6. 381	6. 990	6. 863	6. 4,240	6. 5,015
7. 372	7. 682	7. 917	7. 3,393	7. 2,389
8. 291	8. 995	8. 940	8. 9,033	8. 3,841
9. 764	9. 496	9. 525	9. 7,395	9. 3,099
10. 873	10. 818	10. 814	10. 7,442	10. 8,604

Page 131	Page 132	Page 133	Page 134	Page 135
1. 9,703	1. 37,703	1. 89,277	1. 200,000	1. 975,027
2. 5,946	2. 59,716	2. 86,060	2. 481,158	2. 910,019
3. 9,318	3. 21,120	3. 80,772	3. 788,072	3. 909,577
4. 4,369	4. 91,178	4. 90,971	4. 402,971	4. 565,371
5. 9,105	5. 72,317	5. 91,969	5. 601,991	5. 390,474
6. 8,913	6. 28,227	6. 90,421	6. 290,218	6. 394,801
7. 6,093	7. 55,497	7. 60,107	7. 705,018	7. 823,995
8. 3,689	8. 52,007	8. 60,117	8. 865,021	8. 700,002
9. 8,989	9. 85,092	9. 61,973	9. 404,989	9. 537,867
10. 5,998	10. 31,602	10. 50,110	10. 617,004	10. 879,399

Page 136
1. 6,014,995
2. 8,026,261
3. 4,057,877
4. 2,038,031
5. 5,014,987
6. 8,849,003
7. 2,800,678
8. 7,400,000
9. 4,604,991
10. 2,464,004

Page 138
1. 59,989
2. 68,879
3. 69,757
4. 85,629

Page 139
1. 36,763
2. 38,647
3. 64,899
4. 78,488

Page 140
1. 48,936
2. 63,869
3. 77,033
4. 74,647

Page 141
1. 24,591
2. 79,647
3. 76,998
4. 77,788

Page 142
1. 48,889
2. 69,667
3. 96,979
4. 79,699

Page 145
1. 30
2. 50
3. 40
4. 20
5. 20
6. 40

Page 146
1. 60
2. 30
3. 80
4. 20
5. 90
6. 30

Page 147
1. 50
2. 10
3. 60
4. 30
5. 80
6. 70

Page 149
1. 50
   30
   80
2. 70
   20
   90
3. 30
   40
   70
4. 20
   20
   40
5. 60
   30
   90
6. 60
   30
   90
7. 30
   40
   70
8. 80
   10
   90

Page 150
1. 70
   20
   90
2. 80
   10
   90
3. 50
   20
   70
4. 30
   10
   40
5. 30
   30
   60
6. 70
   20
   90
7. 40
   40
   80
8. 50
   20
   70

Page 151
1. 80
   10
   90
2. 70
   20
   90
3. 60
   30
   90
4. 40
   20
   60
5. 50
   40
   90
6. 90
   10
   100
7. 50
   30
   80
8. 70
   20
   90

Page 154
1. 300
2. 600
3. 600
4. 100
5. 400
6. 800

Page 155
1. 400
2. 300
3. 400
4. 600
5. 300
6. 600

Page 157
1. 400
   500
   900
2. 700
   100
   800
3. 700
   200
   900
4. 300
   500
   800
5. 200
   100
   300
6. 300
   500
   800
7. 500
   300
   800
8. 800
   100
   900

Page 158
1. 700
   200
   900
2. 400
   500
   900
3. 500
   300
   800
4. 800
   100
   900
5. 200
   700
   900
6. 500
   400
   900
7. 600
   300
   900
8. 300
   600
   900

Page 159
1. 300
   300
   600
2. 200
   200
   400
3. 400
   500
   900
4. 300
   300
   600

5. 600
   300
   900
6. 600
   400
   1000
7. 800
   100
   900
8. 300
   100
   400

Page 160
1. 600
   300
   900
2. 900
   100
   1000
3. 600
   400
   1000
4. 200
   700
   900

5. 700
   200
   900
6. 500
   500
   1000
7. 100
   900
   1000
8. 700
   300
   1000

Page 165
1. 48,300
2. 71,900
3. 93,600
4. 23,900

Page 166
1. 87,700
2. 74,300
3. 13,000
4. 35,800

Page 167
1. 91,000
2. 19,000
3. 25,000
4. 46,000

Page 168
1. 36,000
2. 88,000
3. 74,000
4. 95,000

Page 169
1. 50,000
2. 70,000
3. 90,000
4. 20,000

Page 170
1. 36,000
2. 90,000
3. 70,000
4. 20,000

Page 171
1. 400,000
2. 600,000
3. 800,000
4. 100,000

Page 172
1. 400,000
2. 300,000
3. 600,000
4. 800,000

Page 173
1. 7,000,000
2. 5,000,000
3. 6,000,000
4. 3,000,000

Page 174
1. 9,000,000
2. 1,000,000
3. 4,000,000
4. 7,000,000

Page 176
1. 20,000
   10,000
   30,000
2. 40,000
   50,000
   90,000
3. 70,000
   10,000
   80.000
4. 50.000
   40.000
   90.000

Page 177
1. 50.000
   30.000
   80.000
2. 20.000
   40.000
   60.000
3. 70.000
   10.000
   80.000
4. 40.000
   30.000
   70.000

Page 178
1. 600.000
   100.000
   700.000
2. 100.000
   800.000
   900,000
3. 300.000
   500.000
   800.000
4. 700.000
   200.000
   900.000

Page 179
1. 300.000
   600.000
   900.000
2. 500.000
   400.000
   900.000
3. 200.000
   100.000
   300.000
4. 400.000
   500.000
   900.000

Page 180
1. 4.000.000
   3.000.000
   7.000.000
2. 2.000.000
   3.000.000
   5.000.000
3. 4.000.000
   4.000.000
   8.000.000
4. 8.000.000
   1.000.000
   9.000.000

Page 181
1. 2,000,000
   7,000,000
   9,000,000
2. 6,000,000
   3,000,000
   9,000,000
3. 2,000,000
   1,000,000
   2,000,000
4. 4,000,000
   5,000,000
   9,000,000

Page 183
1. 36
2. 40
3. 34
4. 65
5. 90

Page 184
1. 171
2. 866
3. 1,114
4. 898
5. 964

Page 185
1. 1,046
2. 833
3. 1,208
4. 850
5. 915

Page 186
1. 1,671
2. 7,866
3. 9,114
4. 7,898
5. 8,964

Page 187
1. 7046
2. 6833
3. 9208
4. 8750
5. 9815

Page 192
1. $9 + 7 = 16$
2. $24 + 36 = 60$
3. $9 + 4 = 13$
4. $15 + 12 = 27$

Page 194
1. $51 + 37 = 88$
2. $27 + 38 + 2 = 67$
3. $118 + 96 = 214$
4. $41 + 19 = 60$
5. $112 + 97 + 131 = 340$

203

Page 196
1. $48 + 126 + 235 = 409$
2. $51 + 37 = 88$
3. $11 + 16 + 7 = 34$

1. $41 + 36 = 77$
2. $117 + 137 = 254$
3. $37 + 31 = 68$
4. $17 + 12 = 29$
5. $7 + 9 = 16$
6. $12 + 19 + 11 = 42$

www.ingramcontent.com/pod-product-compliance
Lightning Source LLC
LaVergne TN
LVHW081332060426
835513LV00014B/1263